Memoirs of a Rocket Scientist:
From Apollo to Space Shuttle to Minuteman to UAV/BPI

Acknowledgements

I have been lucky to have had the opportunity to work on so many challenging and important problems confronting nearly fifty years of rocketry. From the Apollo space program involving lunar landing, to the Space Shuttle Boosters of the Space Transportation System (STS), to the Minuteman ICBM and its silos, and lately (in "retirement") to the proposed use of Unmanned Aerial Vehicles (UAVs) flying at 20 km altitude to launch high-speed interceptor rockets in defense against attacks by North Korean and Iranian ICBMs during their boost phase.

This career was made very rewarding by the independence and respect given me by my corporate employers (Thiokol Corporation/Morton Thiokol, TRW/Northrop Grumman, and Spectral Sciences Inc), as well as by segments of the Aerospace industry that have asked for my consultation during my retirement.

January 2017

ISBN-13: 978-1537779041

Table of Contents

Introduction

What is a "Rocket Scientist"?

Who Are They?

There is no single definition of who is a rocket scientist. The reason is that a rocket is a very complex system that requires the expertise from a wide range of disciplines: propellant and combustion chemists, physicists, aerodynamicists, mathematicians, structural analysts, guidance and control specialists, material engineers, process engineers, test engineers, mission planners, et al. Most rocket issues are interdisciplinary, and require contributions from all of the above personnel.

What Do They Do?

The first step in developing a new rocket is to identify its mission. Then:

1) The specified mission will constrain its weight, payload, and required performance, i.e. thrust history.
2) Next, trade-offs must be identified, such as how many stages versus system complexity.
3) Then what type of propellant is best for this mission (solid, liquid, hybrid, electric, nuclear?).
4) What type of propellant configuration is optimal in each of these stages (CP, forward fins, aft fins, slots?).
5) What type of igniter is best (pyrogen, pyrotechnic, hypergolic?).
6) What type of nozzle is required (submerged, conical, contoured, plug, extendable?).
7) What kind of thrust control is best (fins, LITVC (liquid-injection thrust vector control), gimbal (pivot)?).
8) What type of case material (metal or filament-wound) and configuration (monolithic or segmented)?

Once those decisions have been made, a detailed design can be made:

1) What is the optimum propellant formulation?
2) Determine the initial propellant shape (solids, hybrids) or oxidizer/fuel ratio (liquids) to generate the required thrust.
3) What will be the resulting combustion chemistry?
4) Predict the resulting flowfield of combustion products in the combustion chamber and nozzle.
5) Predict the resulting structural static and dynamic response of the case, nozzle, and other hardware.
6) Conduct exhaust-plume simulations to identify base heating, flight radiance, and launch dynamics on pad or in silo.
7) Conduct trajectory simulations to ensure that the mission will be achievable.

Finally, the resulting rocket assembly must be validated:

1) Static tests must be conducted to assure that the rocket performance meets the requirements at sea level.
2) Flight tests must be conducted to assure that vehicle acceleration and aero-heating will not cause problems.

The Purpose of These Memoirs

These Memoirs are intended to describe my experiences spanning 40 years in the rocket industry. I discuss the details of many of the problems that I have been tasked to solve, and how those solutions have made a difference to the understanding and modeling of the technology. I envision an audience ranging from anyone who wants to know what a rocket scientist does, to undergraduate college students pursuing studies in engineering and science, to young graduate engineers, to colleagues in the rocket industry. The difficulty that I have faced here was how to make the technology understandable without presenting the complex math and physics required to explain the problems. My goal is to bathe you in it, but not to drown you in it. Hopefully, the result is both entertaining and enlightening. Physics students might enjoy deriving the conservation eqs(5,6) on page 21, chemistry students might want to re-derive the equilibrium equations on page 35, and math students might try deriving eq(3) on page 20, force coefficients on page 74, the range eq(19) on page 87, and several closed-form solutions to rocket problems on pages 92-93. The Anatomy of a Problem (O-Ring) and its Solution is detailed on pages 94-95.

The defining period of my career was my involvement with the Challenger accident, as described on pages 26-32 and 94-95, which led to my nickname "Mr. O-Ring". However, during the remaining 38 years, I was called on to solve a wide variety of problems. My emphasis was modeling ignition and combustion of solid propellants, and the flowfields in and around solid-propellant rockets, but also included the environment in Minuteman ICBM silos (pages 56-57, 76) and developing the first model of automobile gas bag inflator operation (pages 22-23). Despite Microsoft saying "it couldn't be done," I created an Excel Package for engineering plotting on PCs because I needed it (pages 46-48 and most of the plots in these memoirs).

I describe "table-top" experiments (pp 18, 40, 41, 95) that I designed and conducted to help verify assumptions in my models. I demonstrate how easily to make computer programs user-friendly (pages 51, 60-64), and I show how I developed simplified solutions that exposed functional dependence and provided sanity checks on complex solutions; I discovered that it was always best to start simple with modeling and build from there. Chapter 8 presents some important tools to make engineers' tasks much easier. I have continued to consult and publish during "retirement" because the challenges are so stimulating. The technical details of most problems discussed here are presented in my book (Ref 1), for those qualified to access it (page 7).

Many of my colleagues have been smarter and quicker than I, but my bulldog temperament and willingness to work evenings and weekends at home allowed me to make up for my other short-comings. All in all, my successes led to my employers giving me the respect and independence that made it an enjoyable and rewarding career.

My Path to Rocket Science

I didn't plan to become a rocket scientist. It just happened.

As a kid, I loved airplanes, especially military aircraft. I also loved math and history, and was an accomplished clarinet player in high school. I recognized that my best college career path would be to get an engineering degree; I could still enjoy history and play in music groups the rest of my life. So I attended Rensselaer Polytechnic Institute (RPI) and obtained a Bachelor's degree in Aeronautical Engineering (BAe) in 1965.

My parents were very education oriented, so it was understood that I would go on to graduate school. I obtained an MS in Aeronautics and Astronautics from Penn State University in 1967, and a PhD from New York University in 1971, specializing in Fluid Mechanics and Applied Mathematics. After a difficult period of unemployment due to the economy (discussed later), I spent four years conducting Gas Turbine research with Pratt & Whitney in East Hartford, Connecticut.

In late 1976, P&W decided to move 1000 engineers to Florida, but sponsored a job fair for those not willing to move there. I wasn't willing, and Thiokol Corporation invited me to interview in Utah. As a result, I was offered a job as staff scientist supporting the development of the Space Shuttle solid-propellant rocket boosters. It was only then that I became a rocket scientist. The next 40 years were occupied solving rocket problems, 16 years with Thiokol, 16 years with TRW and its acquisition by Northrop Grumman, and 8 years as consultant to the rocket industry.

In those 40 years in the rocket industry I have been tasked with solving possibly the widest variety of rocket problems of any single rocket scientist (so said Air Force Rocket Lab group head Jay Levine). Most of those problems are summarized in my book (Ref 1). However, many problems that occupied much of my time were not included in the book, and are discussed here; for example:

- Development of the first model of automobile gas-bag inflator operation
- Modeling fuel regression in hybrid rockets
- Modeling of the mechanical opening of the 110-ton cover on Minuteman silos
- Modeling the normal thermal environment and a flash fire inside a Minuteman silo
- Modeling a Hot Gas Relief Valve (HGRV) for nozzle thrust vector control
- Modeling and fixing pyrotechnic igniter anomalies
- Modeling the consequence of potential motor detonation during launch of a manned Orion capsule (LAS)
- Modeling the Boost Phase Intercept of enemy ICBMs using UAVs flying at 20 km altitude with interceptors

One purpose of these Memoirs is to describe previously undocumented issues surrounding these problems: how did I get involved, how did I stumble my way to a solution, and what were the consequences of the solution. Some consequences of my analyses were critical: saving gas-bag inflator contracts with the auto industry, restarting the Shuttle booster program or resuming Shuttle flights, verifying the acceptability of tiny lead pellets inadvertently cast into 23 rocket motors. Some were rewarding: an unsolicited phone call from Elon Musk asking me to come work for him at SpaceX after he read my journal article on the connection between the Challenger and Columbia accidents (Ref 3), an invitation from ONERA to lecture them in Paris after they read my journal article on slag generation (Ref 4), invitations to lecture at the Technion in Israel over a 24-year period, requests to lecture at the Von Karman Institute in Brussels (2002) and to ESA in Rome (2015), and many more.

Most of these memoirs are focused on Fluid Mechanics and Combustion, since that was the principal subject of my graduate and subsequent industrial work, although I may be best known around the world for my structural dynamics effort in modeling the deformation and activation of non-linear viscoelastic O-rings. I have also published significant work in the use of strain gages to measure the viscoelastic behavior of propellant and the history of igniter shocks and motor pressurization.

Most of the techniques that I have used in problem solving, as documented here, also apply to many other disciplines: keep an open mind, proceed logically, first formulate the problem as simply as possible, create physical and mathematical models of the problem including simplified analytical solutions, try to obtain or generate experimental data to validate your models, document your work as you go (don't wait for the end of the task), and make code operation user-friendly using menu-driven DOS batch files. Most of the many computer programs that I wrote were aimed at solving specific problems; my tasks were problem solving, not code development. Sometimes, I chose to construct simplified approximate analytical or numerical solutions that were much easier to understand and apply than existing more-sophisticated solutions that were very difficult to understand and utilize. At other times, I was the point man who "cleared the path" or "did the recon" so that others could follow with the more-sophisticated solutions. Rocket science requires lots of math and computer skills.

It has been a very challenging and rewarding career, with many evenings and weekends spent at home doing what there wasn't enough time to accomplish at the office, especially documentation and code debugging. That effort had to fit into my life with wife Alice and raising children Karen and Joshua. They were usually patient and understanding when I disappeared into my home office or stayed up late with my equations and computer programs.

Nonetheless, "all work and no play would have made Mark a dull boy". So I continued my musical and athletic hobbies in parallel. I was principal clarinetist for the Rocky Mountain Symphony in Ogden Utah for many years, and led the Ogden Woodwind Quintet (picture below) for 25 years; my quintet generated a CD titled "Something for Everybody", and played a concert at Ogden's Weber State University titled "Symphonic Favorites for Woodwind Quintet". I also played no-check ice hockey every week with the same Utah group for 32 years, as well as for six years after "retiring" to Sun Lakes, Arizona.

Me on 71st birthday

The Ogden Woodwind Quintet (Me with Clarinet)

During my 32 years of recreational hockey in Utah I got to play with many pros who were playing at least some games in the NHL that same year; they were playing mostly with the Salt Lake Golden Eagles (NHL farm team of the Calgary Flames) and came out to play with us in the weeks before going off to training camp (Rick Lessard, Kevin Guy, Rich Chernomaz, and future New Jersey Devils assistant coach Kurt Kleinendorst). It was fun having an NHL-er as a linemate, and a challenge trying to deke past an NHL-er opponent. After moving to Arizona at age 65, regulars in our hockey group included three former NHL players: Gary Howatt with two Stanley Cup rings with the Islanders, a former Flyer named Granville (I never learned his first name because we all called him Granny), and former Phoenix Coyote Krys Kolanos who was doing rehab for a year after a knee operation, and would play for the Calgary Flames the following year.

After retirement from Northrop Grumman, I taught courses at the Technion (Israel Institute of Technology in Haifa), Redstone Arsenal in Huntsville Alabama, and Orbital Launch Systems in Chandler Arizona. I have also served as consultant for Spectral Sciences Inc., Raytheon, and the University of Texas. I formed the Tempe Wind Quintet during retirement in Arizona; we have played many different venues, including churches, retirement homes, for Chandler Symphony events, and an annual summer recital at a local library.

Rocky Mountain Symphony

James Thomson, Music Director and Conductor

FAMILY CONCERT

GUEST ARTISTS

John Halvorsen, French Horn
Mark Salita, Clarinet
Greta Thomson, narrator

Program

MENDELSSOHN	The Hebrides Overture (Fingal's Cave)	op.26
→ WEBER	Concertino for Clarinet	op.26
	Adagio ma non troppo	
	Piu Lento	
	Allegro	
MOZART	Horn Concerto No. 3	K-447
	Allegro	
	Romanze	
	Allegro	

MINI KINDERKONZERT

MARCH	Washington Post by John Philip Sousa
STORY	The Boy, the Cat and the Magic Fiddle

My Most Important Contributions

Subject	Consequence
Space Shuttle Boosters	
Explained near-fatal ignition overpressure on first Shuttle launch	Identified fix for all future Shuttle launches
Modeled impingement erosion of O-rings on Space Shuttle boosters	Educated Presidential Commission after Challenger
Modeled the pressurization and activation of O-rings in a groove	Tool used throughout rocket and nuclear industries
Explained Challenger and Columbia Accidents	Exposed the criticality of wind shear during launch
Explained error in NASA chamber pressure analysis	Showed safe margin: Shuttle program restarted
Minuteman	
Modeled ignition overpressure in a silo	Explained difference between deep and shallow silo
Modeled nozzle liner ejection	Explained flowfield and thrust consequence
Modeled the physics of opening of a silo cover	Verified ability to open in required time
Modeled melting and vaporization of lead pellets in flowfield	Helped USAF buy-off of 23 contaminated motors
Modeled the anomalous generation of H_2 in silo and resulting fire	Helped identify the cause and safety requirements
User-Friendly Flowfield Modeling	
Created a simple 2-phase CFD chamber flowfield package EVT	Rapid versatile multizone gas/droplet solution
Identified deficiencies and requirements in slag modeling	Provided strong foundation for industry slag work
Derived many closed-form solutions to rocket problems	Analtical solutions expose functional dependence
Created simple vehicle staging analysis and extended to CFD	Provided a sound understanding of staging dynamics
Characterized bi-modal Al_2O_3 particle data from quench bombs	Corrected much misunderstanding in SRM industry
Ran experimental program to model droplet collision/coalescence	Explained droplet transition in nozzle and into plume
Simple models for vehicle lift/drag: Newtonian, blunt body, rarefied	Provided sanity checks on more-sophisticated codes
Improved Operation of Computer Codes	
Created PC plot macros using EXCEL and MATLAB	Free and portable plot package for wide application
Developed menu-driven DOS batch files for 75 computer codes	Enables user-friendly code operation and plotting
Fixed horrible I/O interfaces for many industry codes	CMA92, SPF, ZEUS, NPARC, NARJ now friendly
Developed SUPERBAT GUI to operate all my codes/databases	Northrop Grumman had full access after I retired
Other Unique Models	
First model of operation of gas-bag inflators for automobiles	Saved GM & Ford contracts for Thiokol Corporation
Developed volume-filling models for many unsteady applications	Provided rapid solution and understanding of physics
Wrote rocket trajectory code to validate UAV Boost Phase Intercept	Shows viability of concept for defense from ICBMs
Prolific documentation (in-house, conferences, journals)	My colleagues and industry have benefited greatly

My Book ... My Legacy to the U.S. Rocket Industry

Most of the problems summarized above have been documented in detail in the electronic book (Ref 1) that I wrote as my legacy to the U.S. rocket industry: **Basic Analytical and Numerical Methods for Propulsion and Aerodynamic Analysis of Solid-Propellant Rockets.** The book has been approved for public release by the DOD and Pentagon. However, I have limited its access, as best I can, to the U.S. Government and its contractors, and foreign organizations friendly to the United States (ESA, Israel). I don't want to help Iran, North Korea, China, or Turkey (even though it is a NATO member). The book has been provided on CD in three forms: a text form (522 pages), a slide form (nearly 900 pages), and a highlights form (100 pages). The cover page is reproduced here as page 96.

These Memoirs

A summary of these and other projects are provided in these Memoirs. A bibliography is provided of the primary reports, conference papers, and journal articles documenting the projects discussed herein. Besides my book, my most important articles have been Refs 2-4: the award-winning AIAA conference paper **"Unanticipated Problems and Misunderstood Phenomena in and Around Solid Rockets"**, the article **"Shuttle Disasters: Common Cause?"** in Aerospace America, and **"Deficiencies and Requirements in Modeling of Slag Generation in Solid Rocket Motors"** in the Journal of Propulsion and Power.

<u>Note</u>: The layout of the topics in these Memoirs has been purposely arranged to avoid a page turn in the middle of topic. When a single topic requires two pages, those pages will face the reader when the book is open in order to simplify the presentation. The only violations of this intent are the Reference list, and the reprint of the Aerospace America article. In order to achieve this layout requirement, several topics are slightly out of chronological order due to space limitations.

Hopefully, these Memoirs will be entertaining and enlightening, and will provide guidance on methods of problem solving for other scientists and engineers. As a starting point, I have listed on the next page a set of **"Guidelines for Successful Engineering Analysis"**, based on experience gained from my long technical career. Heed them well.

Guidelines for Successful Engineering Analysis

My nearly 40 years in the rocket business have impressed on me some important guidelines that I recommend for anyone doing scientific or engineering analysis:

1) Conduct an exhaustive **literature search**. Don't try to reinvent the wheel unless the wheel is broken or squeaking. There are many modern search engines like Google or AIAA Electronic Library (www.aiaa.org) to aid this process. The Chemical Propulsion Information and Analysis Center (CPIAC, now CADRE) also provides a literature search service (www.cpiac.jhu.edu).

2) Contact the person(s) who know(s) the **state-of-the-art** for the problem to be solved. If they are helpful to you, be sure to inform them at a later date of any progress you make. By giving as well as receiving, you will have a willing consultant again in the future.

3) Develop a simplified analysis (analytical, numerical, or closed-form) that will identify the **principal mechanisms** and variables of the problem, and will provide a **sanity check** on later more-sophisticated and complex models. Verify that dimensional units match on both sides of all equations.

4) Always search for **experimental data** to verify analytical assumptions. Devise and conduct simple "table-top" experiments if necessary. However, remember that measurements, as well as analysis, may be erroneous.

5) Always run a suite of in-house **test cases** to validate any computer code obtained from outside sources. Don't assume that the code has been validated just because its user's guide shows successful test cases. The supplied test cases may exercise only a portion of the code, may not validate the types of problems of interest to you, may not run on your platform or environment, and won't ensure that you (the user) know how to set up the input or run the code correctly for a new problem.

6) All output and plot files should have at least one **title line** at the top to identify what problem was solved.

7) The variables in all output files should have **dimensional units** explicitly listed. Software vendors have often violated this in the past, saying "they're shown in the manual". The vendors forget that we analysts may use many different tools during a single day, each with different sets of units that easily could be mixed up. A Mars mission (Climate Orbiter on 11/10/1999) failed because of such a misinterpretation (pounds versus Newtons).

8) Always retain the appropriate number of **significant digits** even if the latter ones are zero. For example, the number 7.300 is much more precise than 7.3, which is why EXCEL should be slapped when it drops zeroes (to say nothing of the frustrating column misalignment in EXCEL spreadsheets that results)! Always write a 0 in front of decimal numbers less than 1.0 (for example, 0.30); without it, a .30 could easily be misread as 30 behind an ink spot.

9) **Document** as you go, and do a careful job (you never know how important it may become). Waiting to the end of the project is too late. Having an up-to-date document, even of an unfinished project, serves many purposes:
   a) it provides an immediate status report for management if they ask,
   b) it improves communication among team members throughout the action time of the project,
   c) it forces you to think through the problem early and in a way that you might not otherwise,
   d) it provides a starting point and reminder if you return after being temporarily moved to a different project or return after a prolonged vacation.

   Remember: "**If the analysis wasn't documented, it wasn't done**".

10) There is a bad tendency these days to let **PowerPoint presentations** serve as analysis documentation. This is unwise because the details and rationales of the analyses are often not provided in these presentations.

11) Somewhere in the document, **display all input files** used in the analysis so that all assumptions are visible, and the predictions can be easily repeated or modified in the future. If the input files are very large, at least state their names.

12) User-unfriendliness of computer codes is a major source of frustration and user error. **Menu-Driven DOS Batch Files** are a wonderful way to create a user-friendly interface. They are at least as useful as GUIs, yet much simpler to create, clone, and modify. I can't stress enough the value of Menu-Driven DOS Batch Files (see pages 60ff, 91-92).

13) Modern computational codes are sophisticated and complex. Make sure you **understand the physical meaning** of the predictions; don't just show tables and plots of results. If an engineer presented results that he didn't understand to Thiokol Director Joe Pelham, Joe would throw him out of his office and say "Come back when you can explain their meaning."

I have placed these guidelines at the beginning of my memoirs rather than at the end to ensure that the reader will see and absorb them even if these memoirs are not read completely. They are important. I suggest that you frame this page and hang it on your office wall, maybe surrounded by blinking lights.

Chapter 1 – The Beginnings: El-Hi in the Age of Sputnik

My fourth-grade teacher had a unique way of keeping her class in line. If a student was unruly in class, the teacher would punish him or her by handing them a whole page of addition or multiplication problems that they had to solve. Even though I was rarely unruly, I always requested a page because I loved math. I would treat math and academic challenges pretty much the same way for the rest of my life. Indeed, in retirement I have offered the rocket industry free support on any problems where I have experience, just to stay active in the field and to enjoy the challenge. I guess I haven't changed in 63 years.

The Sputnik Revolution

On October 4, 1957, the Russians launched the first man-made satellite of the earth, named Sputnik. Americans were shocked that the U.S. had not been first to achieve this goal. This shock was compounded when the Soviet Union launched a second satellite, Sputnik 2, on November 3, 1957, and the first American attempt (Navy's Vanguard TV3) failed on the launch pad on December 6, 1957. Finally, the U.S. Army succeeded in launching Explorer 1 atop a Jupiter-C rocket on January 31, 1958. The Space Race was on, and I was then almost 15 years old.

The tangible evidence that the Russians were ahead in the space race created great angst in the U.S. The teaching establishment realized that the U.S. high school curriculum needed to be upgraded to turn out more and better-prepared scientists and engineers if the U.S. was to win the race.

These upgrades had immediate impact on my education in physics and mathematics. The national Physical Science Study Committee (PSSC) quickly developed a new physics textbook for high school seniors that was comparable to textbooks currently used in college. It presented topics such as kinematics and optics that had never been taught in high school before. It was very challenging, but prepared me well for my subsequent undergraduate education in Aeronautical Engineering at Rensselaer Polytechnic Institute (RPI). In addition, upon entering RPI in September 1961, I also benefited from a pair of new textbooks (Physics for Students of Science and Engineering) by Resnick and Halliday that provided a strong foundation for many college students around the country.

Mathematics is critical for engineering and science, and its teaching was also upgraded after Sputnik. The head of the math department at my high school in Valley Stream, New York, was **George Lenchner**. He took three steps to accelerate the math education not only at our high school, but eventually around the country.

Firstly, Mr. Lenchner instituted a special advanced track in mathematics at our high school which allowed qualified students to progress much faster and farther than had previously been possible.

Secondly, the curriculum was revised to start teaching calculus in 9th grade, so that we had four years of calculus by the time we finished high school. Indeed, the freshman year math curriculum at RPI was just a rehash of what I had already learned in high school.

Thirdly, Mr. Lenchner created a math club called **Mathletes** at all the high schools in our district with the aim of competing like physical-sports athletes. Hundreds of challenging math problems in algebra, geometry, trigonometry, and logic were formulated by him and other teachers. Each problem was assigned a time limit for the Mathlete to solve, typically no more than five minutes. Each high school held practice sessions after normal school hours (analogous to sports teams). Then, about a half dozen times a year on a Saturday, teams from all the high schools in the district met to compete. Each school had six competitors for each problem, and each would get a point for each correct answer. The school with the most points won that meet. I was on our "team" during my junior and senior years, and was gratified to find out in later years that Mathletes had expanded across the entire country as **Math Olympiads**. It elevated math nerds to the social status of the jocks. The only difference: there was no roaring crowd rooting for mathletes at competitions.

Incidently, Lenchner was also a CCNY Hall-of-Fame lacrosse player. I asked him how he got to be so good; he said his family was too poor to buy him lacrosse gloves before college, so he developed very fast hands to avoid getting checked. He also had been preparing for a career as a concert pianist until he was wounded leading his Company on D-Day.

My mother as a teen-ager had studied piano under the famous English composer Gustav Holst at Saint Paul's Girl's School in London, and my uncle Henry was conductor of the popular Grant Park Symphony in Chicago. So it's not surprising that I became an amateur musician. I was first-chair clarinetist during most of my high school years and beyond.

I graduated high school in June, 1961, and obtained a summer job calculating the weights of aircraft parts at the Grumman plant on Long Island, within commuting distance of my parents' house. I had been accepted for undergraduate study at RPI, Penn State, and the University of Michigan. Although RPI was by far the most expensive of the three schools, I had a State Merit Scholarship only valid in New York, plus RPI offered me a school partial scholarship. It helped that I had saved earnings from shoveling snow from neighbors' sidewalks in the winters and mowing lawns in the summer. My parents were so supportive of higher education (my mother was a Phi Beta Kappa graduate of the University of Washington during the depression), that they apparently had no second thoughts about any hardship of having to pay the remaining sizable cost of RPI. They eventually reaped their reward when they later got to refer to me as "their son the doctor".

Chapter 2 – Universities and NASA Internships

Rensselaer Polytechnic Institute (1961-1965): The Foundation for What Was to Follow

RPI is a top-rated engineering school in Troy, NY. It was very challenging; indeed, a sizable percentage of freshmen in Aeronautical Engineering flunked out when I was there, in part because they hadn't had the good high school education in physics and math that I had. Also, they had not learned to focus on academics. Nonetheless, my grades through junior year were only fair (3.0 out of 4.0), in part because my study time was in competition with other college activities: I was rushing chairman and steward of my fraternity, played varsity tennis freshman year and varsity lacrosse sophomore through senior years, and played clarinet in the RPI marching and concert bands and in the symphony orchestra (we even recorded an LP album of the Brahms German Requiem with the Simmons College Choir in the acoustically-renowned Bank of Troy Music Hall). The band played in the Field House for all the RPI home ice hockey games, and sometimes on road trips with the team.

The RPI hockey team during my four years there was one of the top teams in the country. The first game of hockey I ever saw was between the RPI alumni and the Swiss Olympic Team (a 3-3 tie). My class had two All-Americans on the RPI team which made it to the NCAA playoffs three of my four years at RPI. Watching this level of ice hockey was so much fun that I began playing on the frozen lakes of Troy with my frat brothers during the winters. I continued to play indoor hockey weekly until my 71st birthday (see picture on page 6).

My senior year, I had to put my nose to the grindstone to improve my GPA so I could get into grad school. I made Dean's List, which helped me to be awarded a Research Assistantship at Penn State for my Master's Degree.

Penn State University (1965-1967): The "Lost" Secret Document

My assistantship at Penn State essentially paid for the work leading to my Master's Thesis on Jet-Flapped Cascades for torpedo control. My office was at the Thomas Garfield Water Tunnel, where I subsequently encountered my first trauma in engineering: dealing with documents classified as Secret.

My thesis was to design, analyze, and test the concept of a Jet-Flapped Cascade for torpedo control. For this work, I was approved for a government Secret Clearance. As a grad student, I was not given a personal safe to store secret documents, so I used the department safe. Whenever I took a secret document from the safe I had to sign it out, and upon returning it, had to sign it back in. One day, another Tunnel employee needed a secret document that I was the last to have signed for, so he asked if he could borrow it. I told him that I had put it back in the safe and signed it in. He said that it wasn't there, and I was the last to have signed it out.

The Security Chief told me I had to find it, because the Tunnel had already had two security violations within the year, and a third would bring in the FBI. I looked everywhere for it, my apartment, my car, my office; it was nowhere to be found. Every lab employee was told to check his safe for the document. If it wasn't found, I might end up with a criminal record and no access to classified material. Lo and behold, the head of the lab found the document in his personal safe, then remembered that he had taken it from the departmental safe without signing it out. Whew, was I (and the rest of the lab) relieved! However, that experience made me turn down all future contact with classified documents unless absolutely necessary.

After finishing my thesis and all required course work, I asked my advisor (Prof. Barnes McCormick) if he thought that I was capable of attaining a PhD. He said yes, so I had to decide how to proceed. At the time, Penn State was focused more on low-speed flow. I felt that I should broaden my graduate education to include high-speed flow. Since New York University had a renowned high-speed Aero and Astro Department, and was located close to the homes of my parents and my then-wife's mother, I applied to NYU and was accepted with an assistantship. However, before starting NYU, I obtained a summer internship with NASA Langley, similar to one I had won the previous summer at the Manned Spacecraft Center (see next).

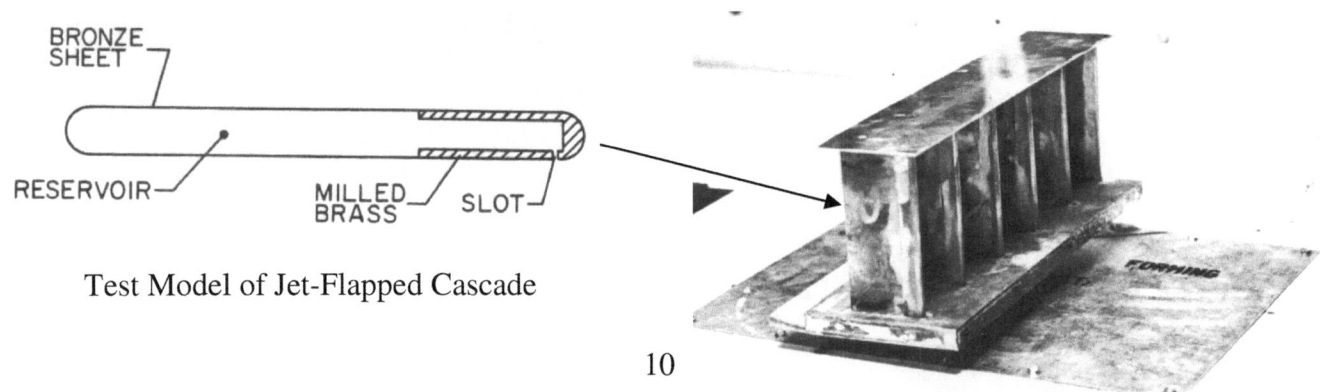

Test Model of Jet-Flapped Cascade

Summer Internships with NASA MSC and Langley (1966-67)

I was lucky to win an internship at NASA's <u>Manned Spacecraft Center</u> (MSC, later renamed the Johnson Space Center (JSC) in Clear Lake, Texas) during the summer of 1966 between my two years at Penn State. I drove down to Pasadena, Texas, after finals were over. At orientation at MSC, I met the other interns, and three of us decided to room together in an apartment complex that allowed us to rent just for the summer. We were assigned to different areas at MSC; I went to work in the Secondary Propulsion Group where we analyzed the future moon landing of the Lunar Excursion Model (LEM). In particular, we were modeling the effect of the LEM descent engine in kicking up dust and rocks during landing on the lunar surface, which could damage the LEM as well as interfere with visual landing capability. I got to sit in a LEM at MSC; boy was it cramped. We also took weekly classes in all disciplines related to the Apollo project, including spaceship design from the famous Max Faget, and rocket nozzle design from G.V.R. Rao (famous for the optimal "Rao Nozzle").

These credentials helped me acquire a second internship at NASA's <u>Langley Research Center</u> the following summer after graduating from Penn State. I worked at the large high-speed wind tunnel, which was run by the famous aircraft designer Richard Whitcomb, creator of the "coke bottle" fuselage shape. My major tasks were data post-processing, and learning to use the CFD codes recently-developed by Allen Vick and Earl Andrews to model rocket exhaust plumes in near-vacuum conditions. Each case ran robustly, but took hours on the then-standard IBM 7090 computer. Thirty-five years later, I tried to type that code into modern computers from their listing in the NASA report (TN D-2327), only to find that modern Fortran compilers viewed much of the 1967 coding as full of errors. This forced me to write my own Method of Characteristics code MOC, which runs in seconds on a PC and works pretty well (although I have yet to implement the tricky coding of the reflection of compression waves from the plume boundary, which made me appreciate the Vick/Andrews success). At the end of the summer, I returned to New York City to begin my PhD program at NYU.

New York University (1967-1971): Tough Grind for the PhD

My assistantship was in the Guggenheim Aeronautical Lab adjacent to the Harlem River in the Bronx, close to the campus of the School of Engineering and Science (years later to be sold to the Bronx Community College, and the Aero Department merged with that from Brooklyn Poly to form the Polytechnic Institute of New York – PINY).

The Guggenheim Lab was run by the famous **Antonio Ferri**, who had escaped Italy during WWII. Three days after the Germans occupied Rome on 10 September 1943, Ferri destroyed the vital equipment at his supersonic research facility at Guidonia and filled a fruit crate with documents of his research before escaping underground. In October 1943, he organized a band of partisans and coordinated attacks on the Germans. After Rome was liberated by the Allies, he made contact with the famous OSS agent Moe Berg (MLB catcher) in order to translate key documents from the crate. Ferri was then brought to Langley to continue his research. Ironically, techs that I used during my internship had also worked for him.

While at Langley, Ferri had developed the Method of Characteristics for solving supersonic flows, which was the basis for the Langley code development mentioned above. My thesis advisor at NYU was Professor Roberto Vaglio-Laurin. He had a very insightful saying: "The equations governing flow in a toilet bowl and flow over a hypersonic wing are identical; only the boundary conditions are different." He of course was referring to the Navier-Stokes equations for viscous flow.

My thesis was to implement/automate his concept that 3D boundary layers on supersonic vehicles could be approximated by a "two-layer model". This model linearized the 3D boundary layer equations using matched asymptotic expansions near the wall ("inner layer") and near the inviscid core flow ("outer layer"), then derived the free constants in these inner and outer solutions by requiring consistency in their overlap domain. I took graduate courses in numerical methods and Fortran (<u>For</u>mula <u>Tran</u>slation) computer programming, necessary skills for engineering analysis. This enabled me to write a computer program to automate the two-layer model. The program was punched into IBM cards; my code filled a large box, which I would submit to the computer center at NYU's Courant Institute in Greenwich Village. Each run would take several hours, during which time I would go out into The Village to browse in the record stores, peek into the jazz clubs, and pick up an occasional hot dog at Nathan's. When the output came back with a list of coding errors, I would identify the fixes, repunch the appropriate cards, resubmit, and head back into The Village. I spent roughly a year commuting daily each way on a pair of subway lines, including Duke Ellington's famous "A-train", from my apartment in the Bronx, through Harlem to the Courant Institute. With today's interactive terminals and rapid throughput, a month's effort in 1970 would probably take a single day today, and I wouldn't even have to leave home.

Vaglio-Laurin was a very difficult advisor, with no empathy for his students. Indeed, when I finally completed my thesis and passed my oral exams in 1971, former students of his who were now employees of the Aero Department or Guggenheim Lab inducted me into the "Vaglio-Laurin Survivor's Club".

We lived in the Bronx for five years, in an apartment building opposite Van Cortland Park (second in size only to Central Park), with alternate-side-of-the-street parking restrictions (we couldn't afford the parking garage under our building). However, it was close to shopping, great restaurants, and only a few blocks from an ice hockey rink where I played with a Fire/Police Department team in a league every week. Played lacrosse some Sundays in the park. Also great tennis; one partner had competed at Wimbledon, another (Mark Rose) subsequently won a National Father/Son tournament.

Chapter 3 – Now What? Gas Turbine Research

Job Hunting for the "Over-Educated" (1971-72): Getting Jobs in Ironic Ways

When I received my PhD in June 1971, the Aerospace Industry was in one of its cyclic depressions. Jobs were few and far between. Since our lives had been on hold for the duration of my thesis work, my then-wife Shelley and I decided to drive around the country visiting relatives, skiing, and job hunting. I played ice hockey at Squaw Valley one evening after skiing; boy, the altitude was a killer after a day on the slopes. I had interviews at several colleges in California and Colorado, but was told that Aerospace Engineers with PhDs were pumping gas for a living. I had an interview with a Vice President of Boeing in Seattle, an acquaintance of my grandfather. It was just after Boeing had decided not to develop a commercial SST, so they were in the midst of layoffs. Consequently, the interview was more of a social visit than a job interview. However, I did get to play ice hockey several times in Seattle.

It seems strange, but I never walked down the aisle at any of my four graduation ceremonies, and only attended two. I was playing in the band for high school and RPI graduations. I was on the way to NASA Langley at the time of Penn State graduation. Without a job at the time of my NYU graduation, I chose not to incur the significant cost to rent the cap and gown for the ceremony at Rockefeller Center, but did go afterward to pick up the ceremonial graduation booklet for posterity.

Upon our return to the Bronx, I started working for some friends who owned a wallpaper factory in Brooklyn, while I continued to job hunt. I used my logic and math skills to straighten out their financial statements, which their previous bookkeeper had simply thrown in a drawer for years. The commute from the Bronx took an hour and a half each way on a sequence of three subway trains, enough time to read the entire daily New York Times, including the ship arrivals.

After about 9 months I had had enough of the wallpaper factory and the commute, so I went on unemployment while continuing to job hunt through the newspapers (remember, there was no internet then, and I was too new to the industry to have any contacts for networking). I interviewed for a job at **Loral** in the Bronx; since the job requirements were entry level, I didn't tell them I had a PhD. But I became worried about that intentional omission since the job required secret clearance, so I didn't pursue the application. Then I saw an ad for a position in Dynamic Meteorology at the **Goddard Space Institute** at Columbia University in upper Manhattan, only a 15-minute commute. They wanted someone with a bachelor's degree in Fluid Mechanics, my specialty. I applied and was invited for an interview. I knew that they would find my PhD over-kill for their opening, but **I thought that if I could teach myself enough about Dynamic Meteorology before the interview (I knew zilch at the time), maybe they would think the fit was too good to worry about my being "over-educated"**. So I went to the NYU library, took out a book on Dynamic Meteorology, and taught myself enough in a week to enable me to sound intelligent on the subject. This preparation seemed to work, since they liked me so much that they invited me back for two more interviews. However, in the end, despite their wanting me for the position, they felt that as soon as a job more-appropriate to my credentials became available I would leave Goddard. So no offer was forthcoming. Very depressing.

After seeing the Glastonbury Connecticut area where some friends lived, we decided that that was the area in which we would love to live. Some time later I was selected for a job interview at Pratt and Whitney Aircraft in East Hartford, next door to Glastonbury. The job related to gasification of coal for use as a fuel in gas turbine engines. **I studied up on the subject before my interview**, as I had for Goddard. After my interview I felt fairly confident that I had made a good impression. Our goal of moving to the Glastonbury area was so close we could taste it. Consequently, we were decimated when I received a rejection letter. I was their second choice, but their first choice had accepted their offer. This was probably the most depressing time of my life.

I took a resume-writing class in Manhattan, run voluntarily by corporate executives at the Advertising Club of New York. I never got a bite from any of the resulting fine-tuned resumes and cover letters that used their corporate recommendations. Ironically, a simple resume that I stuffed in an envelope (even without a cover letter) in response to an ad in the NY Times got me a job with the job-shop **C&D (Consultants and Designers)**. And, joyfully, the job was in East Hartford (next to Glastonbury) supporting Pratt and Whitney, and the salary was good for that day and age ($18,000/year); by comparison, a post-doc in our apartment building became a new assistant professor in Russian Studies at Columbia University starting at $9000/year. Although job-shopping is very transient, offers few fringe benefits like vacation or medical insurance, and your function can be terminated at any time, I moved up to Hartford to start work. We rented a beautiful townhouse in Windsor, just north of East Hartford. I enjoyed driving to C&D that autumn in my second-hand MGA with the top down, through pretty woodland.

A second irony occurred. After working for C&D for several months, I received a phone call from Bob Dring, the head of a Turbine Research Group at Pratt. He told me that he was looking for someone with my credentials, and that the engineer (Harry McDonald) who had interviewed me previously was impressed enough to recommend me. We met for pizza, and he offered me a job in his group (he had to pay off C&D to release me). This reinforced the concept for me that **you shouldn't ever give up on your goals and aspirations**; if you try hard enough, success could be just around the corner.

Within a year, my wife and I had bought a house in Glastonbury. We had made it there after all!

Pratt & Whitney Aircraft (1972-1976): The Famous "Summary"

My biggest contribution to P&W was my document entitled **Summary of Turbine Aerodynamics**. Our Technology and Research (T&R) Group had been having trouble keeping track of the many models for predicting the performance of blade and stator rows in gas turbine engines that power most commercial aircraft. Despite the feelings of my supervisor that it was a waste of time, I compiled a summary of the models, and made as many comparisons as possible among them. In those days before word processors, the entire document had to be typed on a typewriter, figures hand-plotted, and photo-copies made on primitive machines. Consequently, the quality of the copies was pretty marginal.

Nonetheless, the Summary made me famous, although I didn't know it until after I left P&W. Several years later I met a former P&W colleague (Tom Barber) at a Joint Propulsion Conference. He introduced me to several younger guys who worked for him; when they heard my name, they asked if I was the guy who wrote the Summary of Turbine Aerodynamics. Of course, I said I was, and they replied that it was very useful even years later. An even more startling interaction occurred when I was working for TRW in Utah almost 20 years after I wrote the Summary; a TRW employee saw the name on my badge, and asked if I was the guy who had written the Summary at P&W. I said yes and asked how she knew about the Summary. She replied that she and her father had worked at Pratt &Whitney in Florida, and my Summary was his "bible". Wow! **You never know where your publications are going to end up, so do a careful job** (Example #1); this was to occur over and over again during my technical career, as I'll explain later.

We loved Glastonbury Connecticut, rated as an "All-American town". Short commute to P&W (some neighbors rode their bicycles there), easy trip to the opera house or Whaler hockey games in Hartford, beautiful rolling hills, a very educated and affluent community. There was a hockey rink just down the road, with others not far away; I was playing hockey 2 or 3 nights a week. It was an active tennis town with three private clubs, and I was playing the best tennis of my post-varsity career. We lived only 2 miles from I-91, the direct route to the Vermont ski areas, and my wife was a certified Ski Patrol person. It was a great life, but all that changed after three years.

In November 1976, Pratt decided to move 1000 engineers to their Florida facility to work on engines for the military. If a selected engineer chose not to make that move, he would be laid off. However, to ease that process, Pratt held a job fair where recruiters from many technical companies were invited to come interview the engineers who didn't want to go to Florida. I wasn't very happy about the prospect of moving to Florida since it was very humid, hurricane-prone, far from skiing, and at that time there was little ice hockey there. And we were so happy in Glastonbury.

Go West Young Man!

Thiokol Corporation was one of the companies that had a booth at the job fair, and I chose to interview with them. They liked my capabilities and I liked what they had to offer technically and financially, and the opportunity to ski Utah was too good to pass up. At my on-site interview in Utah, I was introduced to an employee who was a former player on the 1954 RPI NCAA championship hockey team, and he assured me that there was a good rink and league 20 miles south in Bountiful. I received a job offer, which I accepted. We put our house on the market, and I moved to Utah in mid-January 1977 to start work. My wife stayed behind to finish packing, sell our house, and to take care of our new-born daughter Karen.

A funny thing happened on the way to begin work at Thiokol. I got off the plane in Salt Lake City, rented a car, and decided to find the Bountiful hockey rink while I was driving past Bountiful on my way to Brigham City, where many Thiokol employees lived. I found the rink and asked at the front desk for any contact info for a senior group. They responded "There's a group in the dressing room right now. Why don't you ask them?" I did, and they asked if I had my gear with me. I said that I had brought everything on the plane except a stick. They said they were short several players, so I could play on their team that night and would lend me a stick and a Pepsi jersey (the team sponsor). So one hour off the plane I was on the ice playing against the BYU hockey club. The altitude (4200 ft) was not much of a problem since I had recently been playing 3 nights a week. It may be hard to believe, but I wore that Pepsi jersey for the rest of my hockey career (43 years); as proof, refer back to the picture on page 6, taken on my 71st birthday (I even scored a goal that day).

I bought a nice house in Pleasant View, a suburb of Ogden, Utah, and moved my family cross-country in March. The commute to Thiokol was 40 miles each way, but it only took 45 minutes in good weather because there wasn't a single traffic light along the way, and half the trip was on a rural section of I-15. Most employees car-pooled, so each drove one day a week plus one additional day a month. It was an easy drive except in the winter snow, and one week each in the spring and fall when the sheep and cattle drives were crossing the road out to the plant. My carpool went in a little late and left work late to avoid traffic (at peak employment, there were 8000 people back and forth every day on the 2-lane road from Brigham City to the plant). Incidently, tourists take that same road to the nearby Golden Spike National Historic Site to watch a daily reenactment of the joining of the transcontinental railroad at Promontory.

Chapter 4 – Rocket Science with Thiokol Corporation (1977-1992)

Rocket Technology Issues

When I joined Thiokol in January 1977, they had government contracts to develop the Space Shuttle Booster, Minuteman III Stage 1, and the C4 Stage 2 Trident submarine-launched ICBM, all solid-propellant motors. That was a fantastic array of programs, funded by NASA, the U.S. Air Force, and the Navy. What an opportunity to learn and contribute! And I did both.

Despite what Prof. Vaglio-Laurin had said about the governing equations being the same, the fluid mechanics of rockets, gas turbines, and torpedoes are very different. Most rockets operate at high temperature (6500°R) and pressure (up to 1800 psi), while gas turbines operate at relatively low temperature (4000°R) and pressure. Torpedoes operate at very low velocities, while rockets can reach Mach 8 in ascent and Mach 20 in reentry. Torpedoes operate in an incompressible medium, liquid-propellant rockets in compressible gas-only media, while solid-propellant rockets generate two-phase (gas/droplet) flow.

First I had to learn the fundamentals of solid-propellant rockets. Unfortunately, there were no good text books on the subject at that time. (When the guys at NASA Headquarters reviewed my book (Ref 1) in 2009, they said they wished they had had such a book at the beginning of their careers. I replied "So did I".) Sutton's Rocket Propulsion Elements, Joseph Foa's (RPI) Elements of Flight Propulsion, and most other text books were primarily academic at that time, and didn't address the real problems in producing a reliable efficient rocket.

Requirements unique to the rocket industry that are rarely taught in university or encountered in other engineering disciplines include
1) flow of high viscosity non-Newtonian liquids (casting (pouring) of solid propellants into motor cases),
2) ignition of solid propellants at up to 10,000 psi/sec,
3) stable and unstable combustion of solid propellants and dynamics of reacting droplets,
4) dynamics and thermochemistry of gas/droplet flows at 6500°R (3500°K) and pressures greater than 1000 psi,
5) structural dynamics with structural/ballistic interaction,
6) external aerodynamics of supersonic/hypersonic bodies of variable cross-section with multiple sets of fins,
7) 6-degree-of-freedom targeting on a rotating non-spheroidal earth,
8) hypersonic flow with gas ionization,
9) radiation and signature from the sun and hot gas/particle exhaust plumes.

Becoming a Rocket Scientist

My only attributes starting work in the rocket industry were a logical mind, facility with applied math and basic physics, and knowledge of Fortran (still useful skill in 2016 since most rocket codes are written in Fortran and few young engineers know the language). I had taken basic propulsion courses at both RPI and Penn State, but they were like crawling before walking.

The best way to learn the fundamentals of solid-propellant rockets was to read the user manuals for the computer programs that more-experienced analysts had written for application to specific rocket issues. These manuals and computer codes provided much important education, but also were deficient in many ways: incomplete documentation, inconsistent nomenclature, poor coding, changing state-of-the-art, and a lack of user friendliness. In order to ensure that I understood certain phenomena, I found it useful to write my own analogous, but simplified computer programs, without the bells and whistles, and then to compare my predictions to those of the legacy codes. This enabled me to verify the assumptions of the original codes, to make improvements to the models, and even to uncover errors in the original codes.

I soon was asked to be Thiokol's representative to the **JANNAF** (**J**oint **A**rmy **N**avy **N**ASA **A**ir **F**orce) Performance Standardization Subcommittee (PSS). That is where I met the guys who were in the trenches, the developers of modern computational codes like OD3P, ODK, TDK, SPP, SPF, et al. That was where state-of-the-art issues of rocket thrust optimization, combustion stability, two-phase flow, and other critical disciplines were addressed. And there were similar JANNAF subcommittees for Nozzles, Structural Analysis, Exhaust Plumes, and other rocket disciplines operating in parallel.

I quickly became a contributor to the field, as I'll discuss below. It became obvious that **the way to solve any technology problem** was to

> 1) First create a physical model of the phenomenon to be studied.
> 2) Then quantify the physical model mathematically.
> 3) Then construct a method to solve the mathematical model.
> 4) Run validation cases by comparison to other analytical prediction methods or to experimental data.
> 5) Finally, display the solution graphically to provide insight and verification.

I documented for the PSS the results of numerous modeling studies (Refs 5-16) that I was conducting, including simulations of propellant casting and ignition, nozzle boundary layers, particle radiation, and motor performance.

During the first decade of my tenure at Thiokol, we had to run computer programs on mainframes (IBM and/or VAX), since PCs were not yet available. Our departments had to pay fees to our Computing Center based on the run times of each program. Computer terminals were available, but non-interactive and only located in special rooms; they only served to allow us to schedule a run. We had to sign up for time on Tektronix terminals to do any graphical plotting, the process was tedious, and plotting software was pretty poor. That was to improve dramatically after the Challenger accident in 1986.

Computational Fluid Mechanics (CFD) was developing rapidly, but during my first decade at Thiokol the commercial CFD codes were expensive, difficult to run, user-unfriendly, structurally limited (e.g., did not allow for body-fitted or multi-zone computational grids, were often unstable), and technical support was often unavailable. Only after the Challenger accident were these limitations addressed: Thiokol engineers now had computer terminals on their desks, commercial codes were rewritten to allow for body-fitted and multi-zone computational grids, and more money became available to run test programs in parallel with code development to validate analytical assumptions, etc.

Not all problems at Thiokol involved rockets. The company's expertise in combustion of solid propellants had wide application. For example, contracts were active to supply flares to the DOD, and gas bag inflators to the automotive industry.

Mythical Rocket Terms

Multiple times during my rocket career, I encountered unique rocket nomenclature (e.g. "**pressure-integral**", "**F over p**", "**max Q**", "**log-normal**"). Often, when I asked experienced colleagues around the industry to tell me the physical significance of the term or how to describe it mathematically, I was surprised to find that they didn't know the answer even though they used the term.

For example, everyone knew that a log-normal distribution is a straight line on log-probability paper. But nobody could tell me how to define the distribution mathematically. In the end, I was able to answer the question and document it (see page 38).

It wasn't until later in my rocket career that I was forced to delve deeper into the significance of some classical terms. For example, two motors with non-eroding nozzle throats and with the same propellant mass and ingredient formulation will have the same pressure integral even if their propellant geometric shapes and resulting pressure histories are totally different.

Likewise, the ratio of motor thrust to chamber pressure F/p is constant during its quasi-steady operation as long as the nozzle throat area is constant. Hence, any increase in F/p during quasi-steady motor burn identifies the degree of throat erosion.

All rocket analysts know that the maximum stress on a vehicle occurs at "max Q" during a flight. But how do you show that mathematically? I called a number of experienced colleagues throughout the industry, to no avail. Then I realized there is a simple explanation: all aerodynamic forces F (drag, pitch, yaw) on a flying vehicle are always written as $F = C_F Q A_{ref}$ where $Q = 0.5 \gamma M_\infty^2 p_\infty$ is the "dynamic pressure". The reference area A_{ref} is a constant, Q is zero at launch (Mach number $M_\infty = 0$), and atmospheric pressure p_∞ and therefore Q is nearly zero at high altitude. At some intermediate time of flight Q maximizes. Thus, forces will also maximize at "max Q" if the force coefficients C_F are constant (which they are not, but their variation is small compared to the orders-of-magnitude variation in p_∞ and M_∞^2).

Misleading Rocket Jargon

There are terms used in the rocket industry that seem oxymoronic. For example, propellant "**grain**" refers to a complete segment of solid propellant, which could weigh thousands of pounds. Likewise, the flow behavior in the solid rocket chamber is called motor "**ballistics**", but has nothing to do with its classical definition of projectiles in flight. (According to Orbital/ATK engineer Arleigh Neunzert, in the early days of rocketry the only people who knew how propellant burned were those with artillery experience. They studied how their grains of powder burned, hence "grain" for the rocket motor propellant.) The study of the projectile was divided into two fields, "internal ballistics" which was what happened inside the gun chamber (another term that translated to rocket chambers), and "external ballistics" or the trajectory of the shell - which is what we now call flight mechanics.)

Documentation

As I stressed earlier, documentation is critical in engineering and science. Not only for technical reasons, but also for the visibility, reputation, and job security of the documenter. I published 181 in-house memos and reports during my 16 years with Thiokol and 144 during my 16 years at TRW/Northrop Grumman. My conference papers and journal articles were later to get me a job twice at TRW, and unsolicited offers to lecture and consult overseas and within the U.S. rocket industry. My motto is "**if the work wasn't documented, it wasn't done**".

In the remainder of this chapter, I will summarize some of the many problems I was called on to solve at Thiokol. However, in order to help readers unfamiliar with solid rockets to understand the issues I had to address, I will first describe some of the basic physics and mathematics involved. Hopefully, enough to bathe you in it, but not so much to drown you in it.

Incidently, much of the graphics have been generated using an EXCEL plot package that I created later while at TRW, and will be described in Chapter 5.

(4A) What Is a Rocket Motor?

What is a Rocket?

A <u>rocket</u> is any type of vehicle that carries all its propellant (fuel and oxidizer) with it, as opposed to air-breathing vehicles that scoop air as their source of oxidizer. <u>Chemical rockets</u> burn a propellant comprised of chemicals, as opposed to ion or nuclear propulsion. Two types of chemical rockets are most common:

1) <u>Solid-propellant</u> rockets use fuel and oxidizer that are pre-mixed together, then poured and cured into a solid mass in a motor case that serves as a combustion chamber when ignited. Solid propellants can be ignited instantaneously even if stored for years, which is why they are required in ICBMs and other battlefield missiles. However, they can't be throttled, and will generate a thrust history that depends on the initial propellant shape. The fuel is typically metal powder (e.g. aluminum) that generates high thrust but a lot of oxide particulate (e.g., Al_2O_3 droplets) that creates both positive and negative effects.

2) <u>Liquid-propellant</u> rockets use liquid fuel and oxidizer that are stored in separate tanks to be mixed and ignited in a separate combustion chamber. Liquid propellants are volatile, but are used when thrust throttling is required.

A third type of chemical rocket (called a "<u>hybrid</u>") consists of a liquid oxidizer and a solid fuel, and has been the subject of significant research in recent years; however, despite its superior safety and environmental characteristics, it has yet to find a practical commercial niche due to performance limitations, and will be discussed minimally here (see page 37). Other types of propulsion (nuclear, ion, MHD) utilize other technologies with which I was never involved.

Although the focus of this memoir is on solid-propellant rockets, it should be noted that most of the physics and many of the analytical and numerical methods discussed here (with the particulate removed) are also applicable to liquid and hybrid rockets: combustion, nozzle flow, plume flow, external aerodynamics, heat transfer, launch issues, and trajectory analysis.

Rocket Motor = Case + Propellant + Igniter + Combustion Chamber + Nozzle

A rocket motor consists primarily of five components (figures below): (1) a strong insulated **case** that contains (2) **propellant** that is either (a) premixed solid fuel and oxidizer or (b) liquid propellant in separate volumes of fuel and oxidizer under high pressure, (3) which is **ignited** chemically (liquids) or by convective/radiative/pyrotechnic heating (solids), (4) and burns in a **combustion chamber** that is either (a) the "void volume" enclosed by the solid propellant or (b) a separate pressure vessel where the liquid fuel and oxidizer mix and ignite. (5) The resulting products of combustion are exhausted through a **nozzle** which creates the motor thrust (a) by expelling gas/droplet <u>mass</u> (action/reaction) and (b) from high gas <u>pressure</u> pushing forward along the nozzle wall (see eq(4) on page 20), which explains why rockets can be propelled through the vacuum of space.

Schematics of Typical Rocket Motors: (a) Solid Booster, (b) Solid Space Motor, and (c) Liquid Engine

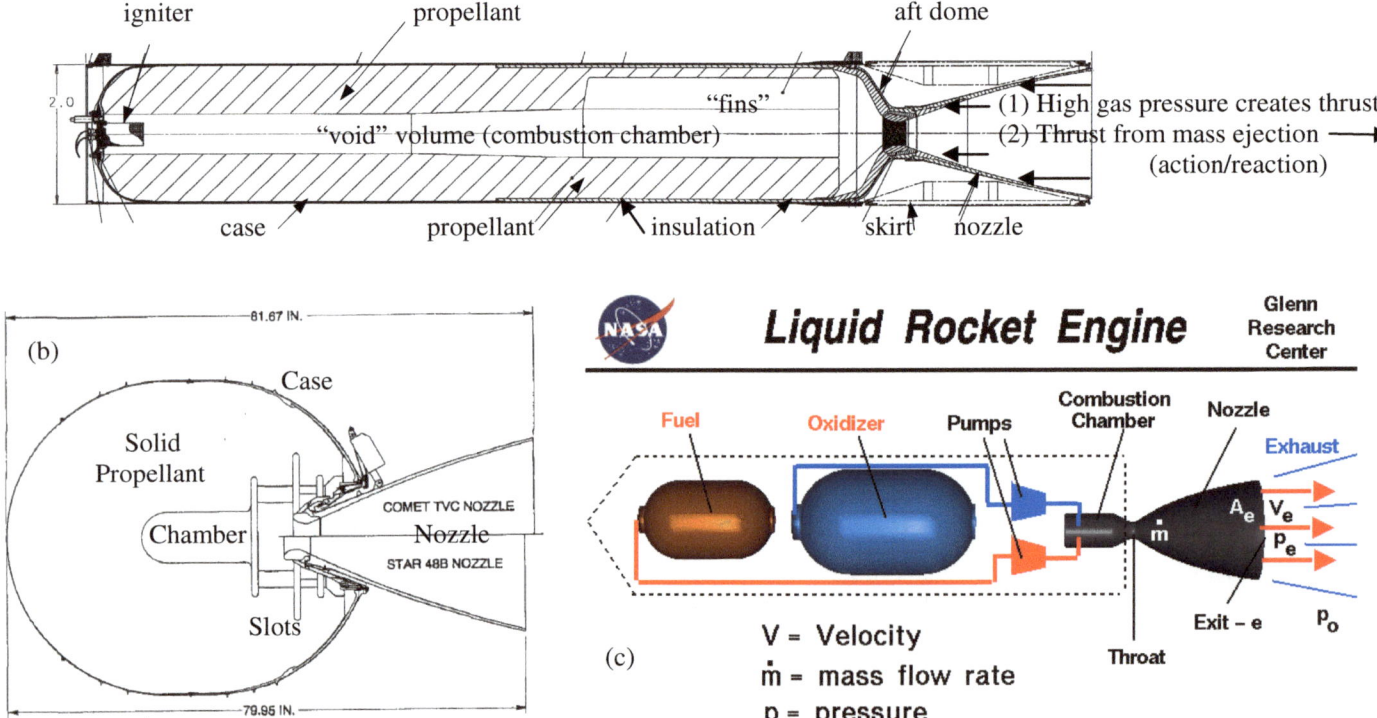

Propellant is comprised of two components (fuel and oxidizer). Examples are shown in the table below.

Solid propellant: Fuel, oxidizer, and binder are mixed together, then cast and cured into a single unified mass
Liquid propellant: Fuel and oxidizer are stored as separate liquid constituents, then spray-mixed
Hybrid propellant: Fuel is solid while oxidizer is gas or gasified liquid that flows over the solid grain

The fuel in solid propellants is usually metal powder because burning it greatly increases the temperature of the combustion products, and therefore the motor thrust. Typical solid rocket motors (SRMs) contain 15-20% aluminum powder, which burns to form liquid aluminum oxide (Al_2O_3) droplets that comprise about 25-35% of the product mass fraction.

Solid Propellants		Liquid Propellants	
Fuels	Oxidizers	Fuels	Oxidizers
Al	AP	RP-1	H_2O_2
B	HMX	H_2	O_2
Mg	RDX	UDMH	N_2O_4

Igniter

Igniters in solid rocket motors (SRMs) heat the main propellant to ignition within milliseconds by exhausting hot gas/droplet flow; they are of two types: either (1) a mini rocket motor ("pyrogen") at the headend of the chamber whose nozzle is replaced by a set of outflow orifices (see page 57), or (2) a "pyrotechnic" igniter consisting of a basket of B/KNO_3 pellets or a magnesium/teflon mix (see page 77). The pyrotechnic igniter also spews hot condensed particles onto the main propellant.

Rocket propellants are highly reactive materials. So it is no surprise that igniting them causes a severe environment. Indeed, most rocket failures occur at propellant ignition. Examples presented in these memoirs include the first launch of Vanguard TV3 in 1957, the burn-through of O-rings at ignition of a Shuttle booster, and the first static firing of Titan SRMU that exploded on the test stand. Add to those accidents, the Takata gas-bag inflator failures at ignition of its ammonium nitrate gas-generant. The first Shuttle launch (STS-1) nearly failed due to the resulting quasi-shock wave generated at booster ignition (see page 25). Vehicle staging occurs at upper stage ignition, and sometimes results in mission failure.

My wife Alice has experienced the power of rocket ignition. When Shuttle boosters were test fired at Thiokol's Promontory test site, their nozzles faced <u>east</u> toward Brigham City and Ogden. The sonic boom of the motor ignition would travel across the desert toward civilization. It would suddenly loudly rattle the windows at the library 40 miles away where my wife was a volunteer. At first, she thought it was an earthquake (the Wasatch front is prone to them), then realized it must have been a booster firing, and she sometimes called me to confirm it. On the other hand, I often stayed in my office several miles <u>west</u> of the test site during a test (I didn't need to watch every test), and I didn't even hear or feel the firing. That directional ignition wave was to play an important role in Shuttle launches at the Cape (see page 25).

Combustion Chamber

Provides a void volume in which propellant burns to convert chemical potential energy to thermal potential energy
Solid rocket: Combustion (void) volume is enclosed by propellant mass for much of the burn time
Liquid rocket: Combustion volume is where fuel/oxidizer sprays mix and burn
Hybrid rocket: Combustion volume is a set of axial ports through the solid fuel (see page 37)

A solid-propellant booster can be comprised of a single segment, or multiple segments pinned or bolted together, with propellant "grains" separated by slots that are typically "inhibited" from burning on one side (e.g., the Shuttle booster). The rubber inhibitors don't burn back as fast as the propellant, so they stick out into the bore flow as the propellant burns back, thereby shedding vortices (a major problem for Ariane 5). Some of the larger Al_2O_3 droplets are trapped behind submerged nozzles to form slag pools (as much as 4500 lbm (2000 kg) in Titan SRMU and Ariane 5).

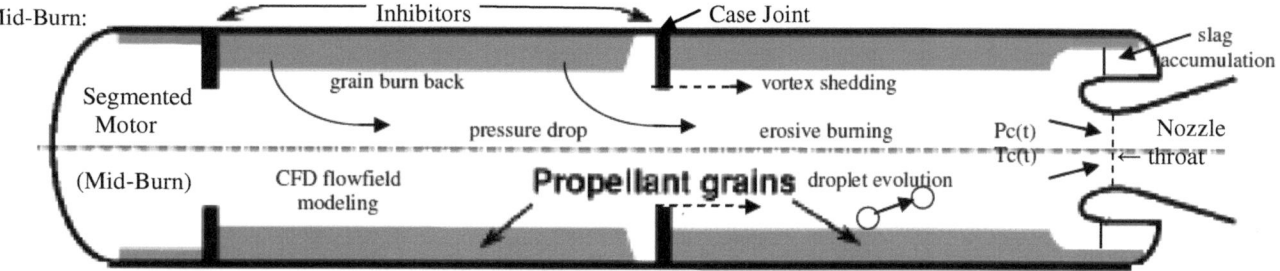

Nozzle (Solid, Liquid, Hybrid Motors)
Converts thermal potential energy of the gas/droplet flow into kinetic energy in the form of thrust. The nozzle may be submerged into the chamber, and has a minimum flow area at the "throat" where the flow is "choked" (Mach 1). The flow velocity increases along the conical or contoured expansion section of the nozzle, while pressure and temperature decrease.

Multi-Stage Vehicles:
Single-segment motors are sufficient for short-range or small payload missiles. Multi-segment motors are used for missiles requiring larger payloads or long range (Shuttle). Alternately, multiple single-segment motors can be stacked in line (e.g. ICBMs, see pages 24, 58) or clustered by "strapping" a bunch of single-stage motors around the core vehicle (see back cover) to form a multi-stage missile for even longer ranges. Only the lowest stage is burning at one time and can be dropped when the propellant is fully expended, which eliminates dead weight and increases vehicle performance.

(4B) Casting and Burnback of Solid Propellants

Modeling the Casting of Solid Propellants With High-Viscosity Resin and CFD

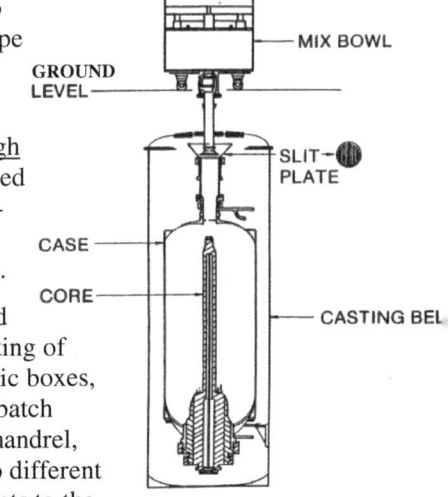

The ingredients of solid propellants are mixed in a huge mix bowl, then poured ("cast") into the motor case. A "mandrel" (core) sits along the center of the case and creates the void shape of the combustion chamber by excluding the required propellant volume. Multiple batches of propellant must often be poured. The motor is then heated and cured, like baking a cake.

The Problem: A static test firing of a full-scale booster motor resulted in a <u>near burn-through of the motor case</u>. The production line was stopped, until we could explain why that happened and fix it. A full-scale motor was cast with alternate propellant batches colored with carbon-black to reveal the batch interfaces, along which high burn-rate binder would accumulate. The cured motor was cut open for viewing the propellant (very expensive, see photo below).

My Solution: To gain a better understanding of these interfaces more cheaply, I devised and ran a "table-top" experimental program using clear high-viscosity resins to simulate the casting of solid propellant (Refs 14,17). Our lab manufactured a bunch of small rectangular clear plastic boxes, open only at the top, into which we poured a clear batch of resin. We then poured a second batch of the resin which had been dyed red, down one wall of the box that simulated the casting mandrel, while videotaping the process from the side. Resins of different viscosities were poured into different boxes. In some tests, my supervisor Dwight Clark added tiny plastic (massless) colored pellets to the poured resin so we could see the flow streamlines. Results for the right half-plane are shown below:

Pouring Resin Down a Wall to Simulate Mandrel Casting of Multi-Batch Solid Propellants

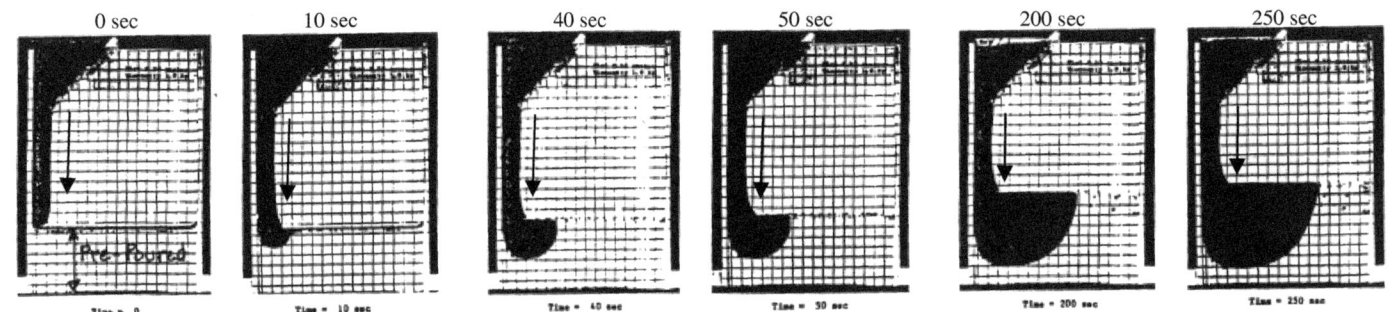

I also modified an existing Los Alamos CFD computer program to predict analytically the flow of the propellant during casting (Ref 18). Similarity of the resin casting and CFD results to the interface shapes in the full-scale motor was quite good:

Numerical Simulation of Resin Flow into Existing Batch Using Modified Sola-VOF Code   Motor Cross-Section

These results suggested a fix to the problem: the "slit plate" through which the propellant was poured into the motor case had to be redesigned from parallel slits to radial slits to force axisymmetric interfaces between cast propellant batches. This would allow predictability of the location where the batch interfaces containing higher propellant burn rate intersected the case wall, so the case insulation could be thickened there.

Modeling Burnback of Solid Propellants

Once the cured propellant surface is ignited, it burns back perpendicularly until it is (hopefully) all consumed. Thus the initial shape of the void volume determines how the burnback proceeds, and therefore determines the history of pressure inside the combustion chamber. The initial propellant surface shape must be chosen very carefully in order that the combustion process generate the desired motor pressure (and therefore thrust) history (see page 20), and that the propellant burn in a structurally-sound manner (the right hardness with no fractures). Several examples of propellant burnback contours are shown below:

Endburning Gas Generator

Star-48 Space Motor with Radial Slots

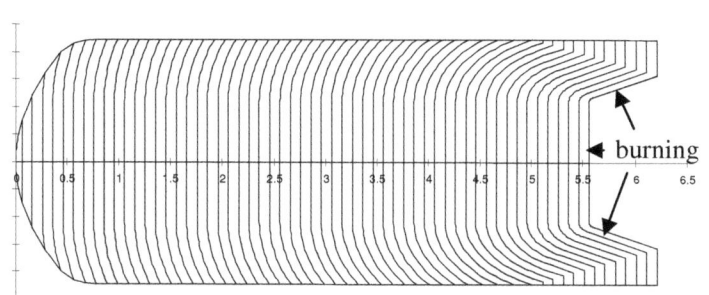

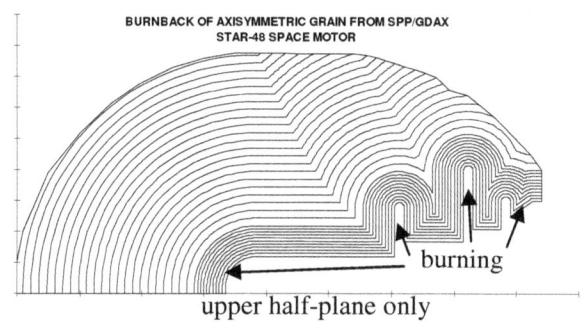

upper half-plane only

Endburning bores burn back axisymmetrically. The **center-burning** Star-48 space motor shown above has three radial slots that are machined in the shape of toroids, consequently the bore is also axisymmetric. However, in some motors, the slots have to be designed as three-dimensional "fins" in order to create sufficient burning surface area **S** and chamber pressure p_c early in the motor action time. The fins consist of anywhere from 3-11 radial arms, so their cross-sections look like **stars**.

Much of the solid-propellant rocket industry uses the computer code SPP to simulate the burnback of the propellant. An example is shown below for the forward segment of the Space Shuttle booster. The bore in the aft half of this segment is nearly circular-cylindrical, so the burnback proceeds nearly cylindrically, except for a small region at the aft end that is partially inhibited. However, the forward half of this forward segment consists of an N=11-arm "fin", i.e. star:

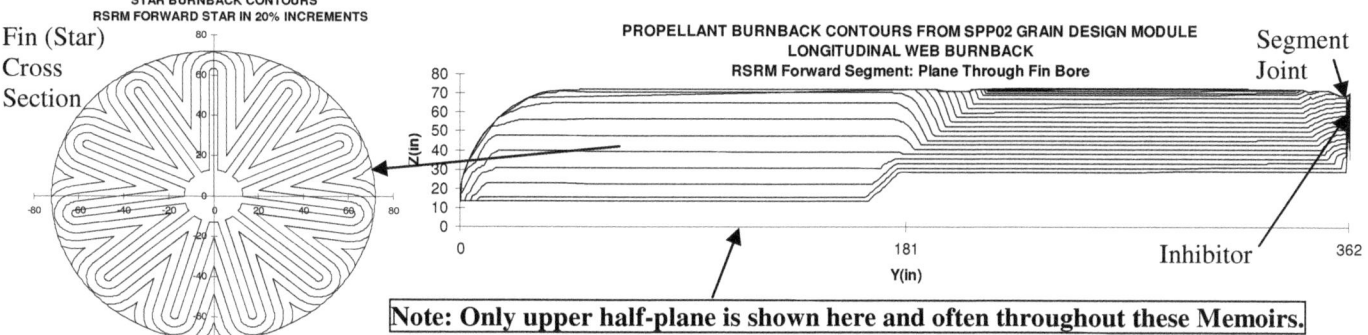

Note: Only upper half-plane is shown here and often throughout these Memoirs.

SPP calculates fin burnback numerically, and can't plot the resulting contours as continuous lines. To validate the SPP burnback contours for axially-uniform fins, I wrote a computer program **STAR** to construct and plot the burnback <u>purely algebraically</u> by focusing on a unit pie-segment, which together with its mirror image formed each of the N arms of the fin (Ref 19). Examples are shown above for an 11-arm fin, and below for a fin with four arms with radially untapered slots, as well as one with 7 radially-tapered slots:

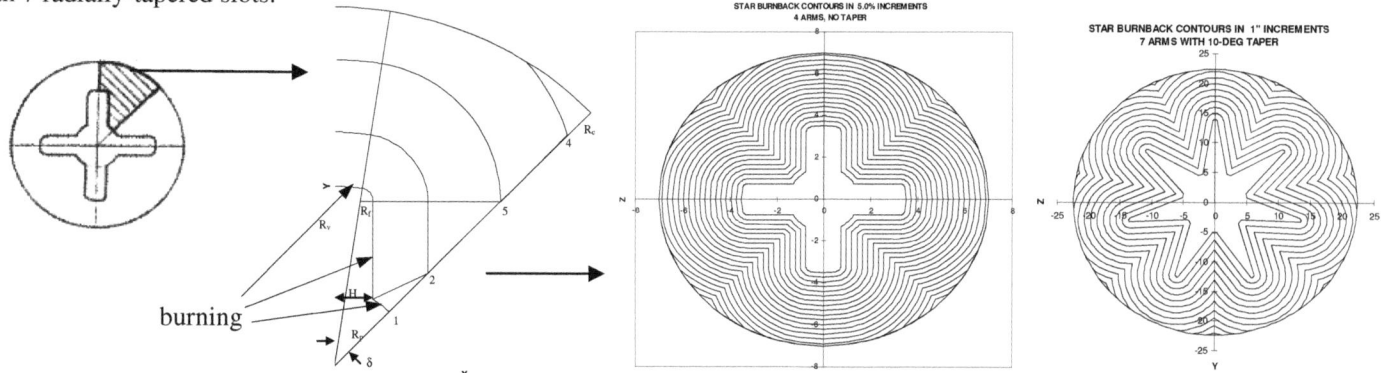

STAR calculates the burning perimeter exactly at any desired burnback distance **b**, and therefore the burning surface area **S(b)** of a <u>planar</u> star (whose cross-section is constant in the axial direction), and the "port" (bore) cross-sectional area of the combustion chamber. STAR was used numerous times to verify SPP numerical solutions, and to visualize the burnback process, as well as to calculate S(b) for truly planar stars, like that in the Minuteman Shroud Tractor Motor.

Ballistician's Equation: Single Most-Useful Equation in Solid-Propellant Rocket Science

Warning: Math Ahead !

The pressure history $p_c(t)$ in the combustion chamber can be approximated explicitly for **quasi-steady** burning at time t by equating the mass flow rate $\dot{m}_{surf}$ from the burning surface to that $\dot{m}_{nozz}$ passing out through the nozzle throat of area A*:

$$\dot{m}_{in}(t) = \dot{m}_{surf} = S(b)\,\rho_p\,\dot{r}(t) \qquad (1)$$

$$\dot{m}_{out}(t) = \dot{m}_{nozz} = \frac{p_c A^* \psi^*}{C^*} \qquad (2)$$

where

- $S(b)$ = burning surface area at burn back distance b
- ρ_p = propellant mass density $\qquad (\dot{\ }) $ = rate d/dt
- $\dot{r}(t) = \dot{r}_{ref}[p_c(t)/p_{ref}]^n\,\phi$ = propellant burn rate at time t
- C^* = propellant characteristic velocity (combustion parameter)
- ψ^* = 1 for a choked throat, <1 if unchoked
- $\phi = \exp[(1-n)\pi_k(T_o - T_{ref})]$ = effect of propellant initial temperature T_o on burn rate

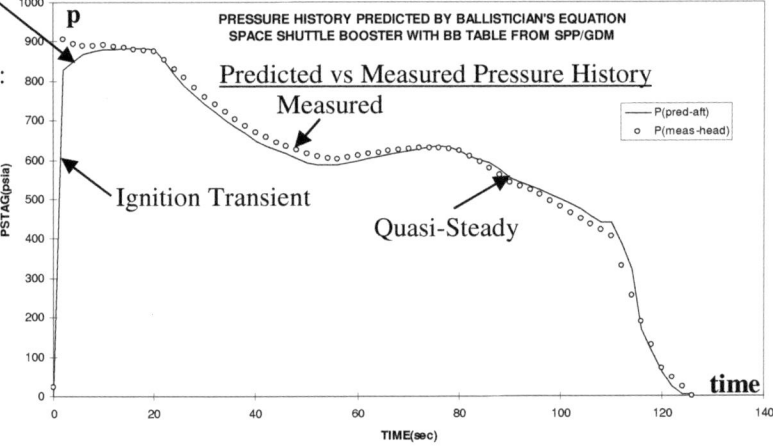

Mass flow-rate balance in rocket chamber

For quasi-steady operation (valid for all burn times except the rapid ignition transient), the equality of these mass flow rates yields an explicit equation for the history of chamber pressure:

$$\dot{m}_{surf} = \dot{m}_{nozz} \quad \rightarrow \quad \boxed{p_c(t) = p_{ref}\left[\frac{S(b)}{A^*}\frac{\rho_p\dot{r}_{ref}C^*\phi}{p_{ref}}\right]^{1/(1-n)}} \quad \text{where} \quad b(t) = \int_0^t \dot{r}\,dt \qquad (3)$$

This equation is even quite accurate for most unsteady motor operation, other than the initial ignition transient:

Eq(3) is so important, that I automated its evaluation as computer program BALLIST with special bells and whistles, such as including bi-propellants, and solving the inverse problem of deducing the burn-back history $S(b)$ given the pressure history $p_c(t)$.

Now that chamber pressure and $\dot{m}_{nozz}$ are known, an approximation of the quasi-steady thrust is given by

$$\boxed{F = \dot{m}_{nozz}\,u_e + (p_e - p_{atm})A_e} \qquad (4)$$

where subscript "e" refers to properties at nozzle exit, and "atm" is atmospheric.

PRESSURE HISTORY PREDICTED BY BALLISTICIAN'S EQUATION
SPACE SHUTTLE BOOSTER WITH BB TABLE FROM SPP/GDM

Predicted vs Measured Pressure History
Measured
Ignition Transient
Quasi-Steady

The Solid Propellant Performance Program (SPP)

Calculation of the propellant burnback $S(b)$ is made by any number of computer codes in the industry, but the gold standard is **SPP** (**S**olid **P**ropellant **P**erformance **P**rediction Program, which should probably be called SPPPP). The burnback contours shown on page 19 for the Shuttle booster (RSRM) first segment and the Star48 space motor were generated by SPP.

SPP also calculates 1D (One-Dimensional) chamber ballistics, combustion stability, equilibrium and kinetic chamber and nozzle thermochemistry, nozzle gas/droplet inviscid flowfields, wall boundary layers, and of course motor thrust in the form of specific impulse I_{sp}. The ideal I_{sp} is debited by various losses, which are predicted using sophisticated physical and mathematical models in SPP. The code is used throughout the U.S. rocket industry; export versions are also used by our NATO allies. Appendix H of my book (Ref 1) describes each of the performance loss mechanisms in simplified analytical terms.

I subsequently wrote a Post-Processor named SPPPP (probably should be called SPPPPPP) for both the 2002 and 2012 versions of SPP that rewrites

(1) the output surface area table $S(b)$ into the form needed by BALLIST,
(2) the propellant burnback contour coordinates into the form required by my plot package,
(3) the output nozzle exit flow into the form needed by my user-friendly version of the plume code SPF-III, and
(4) the boundary layer edge conditions and wall friction and heating coefficients in tabular/plot-able form.

SPP has been developed by Software and Engineering Associates (SEA) of Carson City, Nevada. We have consulted back and forth for nearly 40 years on solid rocket technology. President Doug Coats has been very gracious in helping me with SPP user issues, and I have given them several of my codes (VOLFIL, OD3P). They patterned their Fully-Coupled Transonic nozzle flow module after my two-phase extension NAP2P of Mike Cline's gas-only Nozzle Analysis Program NAP. Doug is a font of knowledge in the rocket industry, and a good friend.

(4C) The Great Utility of Lumped-Parameter Analysis

Unsteady Volume Filling Analysis

The Problem: Flow in and around a rocket motor is very complex, both geometrically and phenomenologically. In order to begin to understand the dominant physical mechanisms, it is useful to construct as simple a flow model as possible. Such a model is called a "volume-filling" or "lumped parameter" analysis, which predicts the average spatially-uniform ("lumped") but unsteady properties in a defined volume with inflow and outflow. Previous models only solved for pressure history.

Warning: Math Ahead !

My Solution: One of my first contributions was to create a "volume-filling" code **VOLFIL** that was more general than had been previously available (Ref 20). In this model, both the mass _and_ energy conservation equations for volume V are solved for the histories of its stagnation pressure p and temperature T. For a calorically-perfect inert ideal gas $p=\rho RT$ with mass $m=\rho V$ and energy mC_vT ($C_v=C_p-R$, $\gamma=C_p/C_v$):

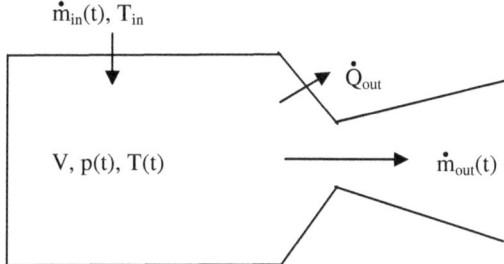

$$\frac{dp}{dt}=\frac{RT}{V}\left[\dot{m}_{in}-\dot{m}_{out}\right]+\frac{p}{T}\frac{dT}{dt}-\frac{p}{V}\frac{dV}{dt} \qquad (5)$$

$$\frac{dT}{dt}=\frac{RT}{pV}\left[\gamma(\dot{m}_{in}T_{in}-\dot{m}_{out}T)-(\dot{m}_{in}-\dot{m}_{out})T\right]-\frac{RT}{VC_v}\frac{dV}{dt}-\frac{\dot{Q}_{out}}{C_v} \qquad (6)$$

Mass flow rates $\dot{m}$ in and out of the volume are specified or calculated. $\dot{Q}_{out}$ is rate of heat loss through the volume boundary. If the volume-filling model applies to a burning solid-propellant rocket, the mass flow rates are calculated using eqs(1,2) from the previous page. By solving the temperature equation, I was able to add models for convective and radiative heat flux into the propellant, then calculate the resulting thermal wave into the propellant, and subsequent propellant ignition. There is no momentum equation to calculate velocities or pressure waves in the volume; those must be modeled using Computational Fluid Dynamics (CFD) codes like SHARPIT (see page 57).

I subsequently used the VOLFIL methodology over and over again to model (1) propellant ignition in rocket chambers, (2) vehicle staging, (3) gas-bag inflator operation, (4) pressurization and jet-impingement erosion of O-rings in Shuttle booster joints, (5) operation of gas generators, and many other phenomena. Colleagues were very impressed at how quickly a first-order solution could be constructed using this methodology. Some examples are shown below.

Wet Grain Ignition Modeling using VOLFIL

The Problem: Observers of the first canister steam-launch of a Peacekeeper ICBM were so shocked at the extended ignition delay of the ejected vehicle that they at first feared a hang-fire (non-ignition) was occurring and the motor would fall back to earth. They were relieved when the motor did ignite. However, they hadn't realized that during steam launch, steam that is injected into the bore condenses on the cold surface of the propellant as a film of water. Before the propellant can be heated to ignition, this water must first be vaporized.

Our Solution: As a result, a horizontal static test of a Peacekeeper missile was devised whereby a hose on a boom extending into the motor through the nozzle would spray water onto the propellant surface immediately before ignition. (The dolly holding the boom would then be blown over the hill after motor ignition.) Given the thickness of the water film measured in several inert spray tests, I was asked to modify my volume-filling code **VOLFIL** to simulate the presence and subsequent vaporization of the water film.

I modified the ignition model in VOLFIL to require the igniter products to raise the water to its saturation pressure and vaporization temperature, then to vaporize the film, and only then to preheat the propellant. I predicted the resulting ignition transient for the upcoming static test (Ref 21). As it turned out, the measured transient matched my predictions closely (see figure). After the test, the Director of Engineering (John Thirkill) came into my office to congratulate me, and to tell me that I would be given a bonus! It always feels good to be recognized and rewarded.

A Lesson: A national survey taken about that time revealed that **the most important factors for job satisfaction** among engineers and scientists were **recognition, independence, and work environment**; I think that salary was only fourth most important. Indeed, I was subsequently given independence and recognition not only at Thiokol, but also later at TRW and Northrop Grumman. That was highly motivating, and turned me into a workaholic.

21

Modeling Gas-Bag Inflator Operation Using Multi-Chamber Volume-Filling (1980)

Thiokol was in the forefront in developing gas-bag inflators for automobiles in the early 1980s because inflators have similar requirements as solid-propellant rockets: gas generant (propellant) that will remain inert and stable, and ignite rapidly on demand for decades in a car, or ICBM in a silo. Consequently, the original inflators burned rocket solid propellants to generate gas (mostly nitrogen, which is why we call them gas bags rather than air bags) that would rapidly fill and inflate a bag.

The Problem: The customers (Ford and GM) were threatening to terminate their funding of the gas-bag project unless Thiokol developed a model of inflator operation that could quantify the rate of gas generation and pressurization as a function of the inflator composition and configuration. Too many test anomalies were occurring from excessive inflator operation. (For example, a common story of the time was that sometimes test inflators fired so fast that passenger dummies were blown through the back window of the car.) However, the analyst initially assigned that task was unable to create a viable model.

My Solution: I was then asked to create a model, which I did successfully using a volume-filling approach (Ref 22). The inflator was modeled as a sequence of coupled volumes (see below) with outflow from one providing inflow to the next. A sample solution is shown below (sorry for the quality of the graphics, but these are from my 35-year old archives). The customers were happy, and <u>Thiokol's very profitable corner on the market was retained</u>.

The profitability of Thiokol's gas-bag business prompted the Morton-Norwich company to merge with Thiokol in 1982 to form Morton-Thiokol Incorporated. However, Morton later took the gas-bag business when they spun off from Thiokol.

<u>Physical Model</u>

Cross-Section of Early
Gas-Bag Inflator

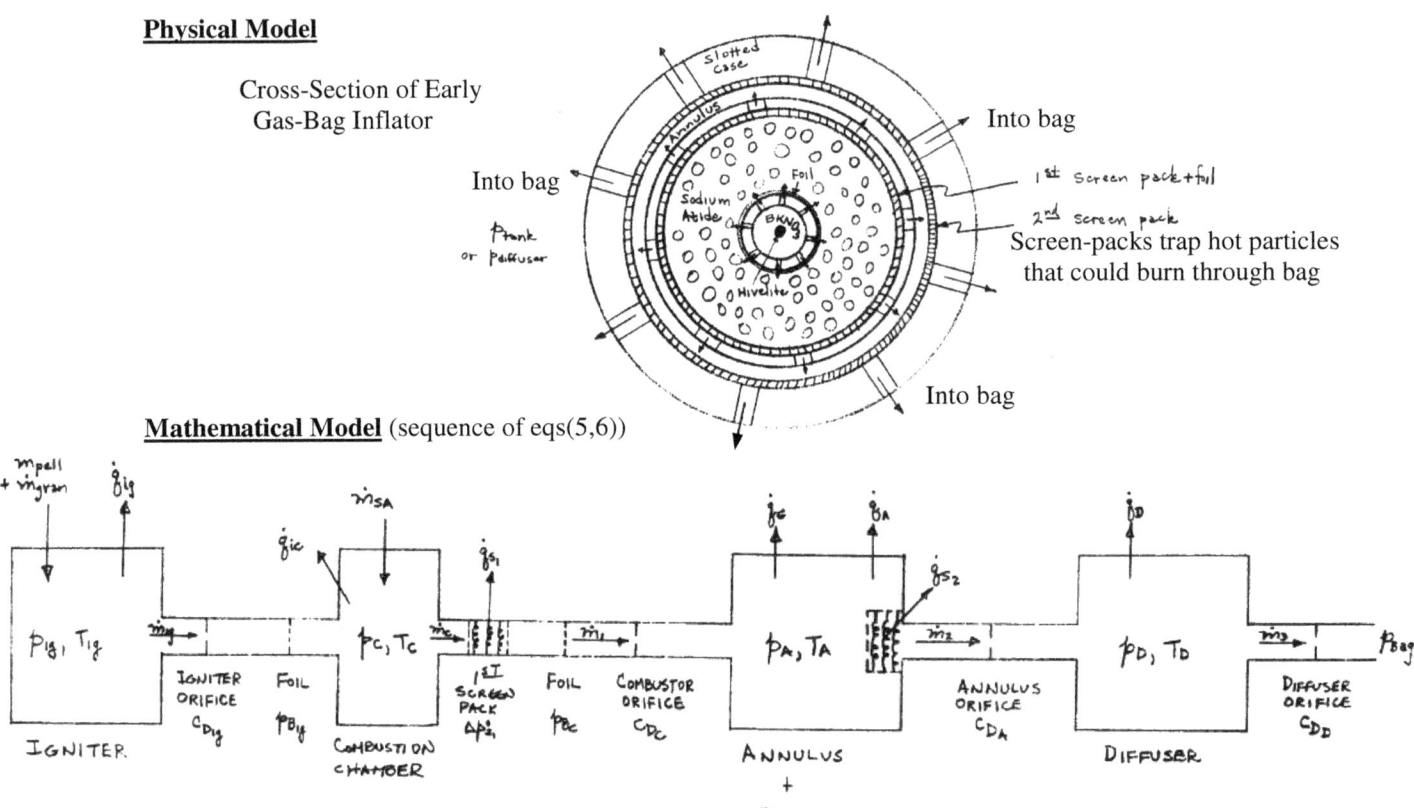

Into bag

Into bag

1st Screen pack + foil

2nd screen pack

Screen-packs trap hot particles
that could burn through bag

Into bag

<u>Mathematical Model</u> (sequence of eqs(5,6))

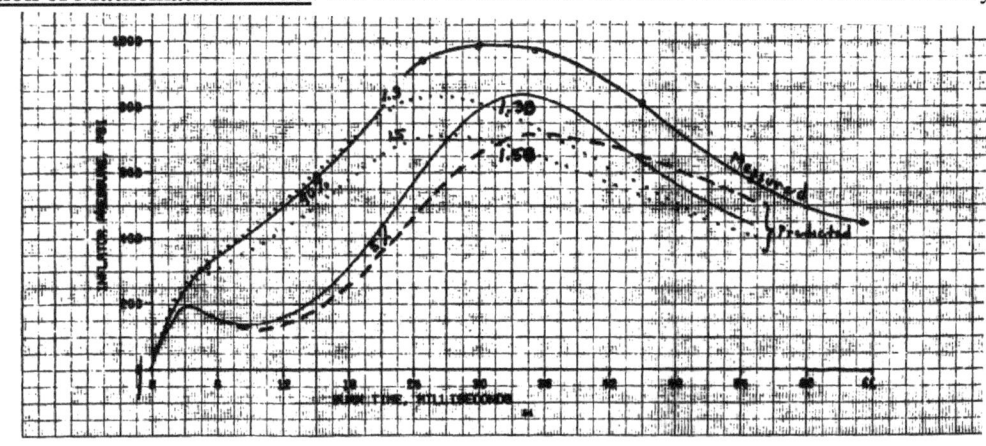

<u>Solution of Mathematical Model</u>: The Effect of Generant Burn Rate on Inflator Pressure History is Shown Here:

Pin-Hole Leak in Gas-Bag Inflators (1985): More Volume-Filling and a Clue for Takata in 2015

The Problem: The Quality Assurance (QA) people at the gas-bag inflator manufacturing plant in Ogden, Utah, found that one of the welds had a pin-hole leak. This could allow the ingestion of moisture into the inflator during a diurnal cycle:

1) If an automobile sits in the sun on a hot humid day, the air in the constant-volume inflator will get hot, and by Boyle's Law ($p=\rho RT$) its pressure will rise, and some air will be exhausted slowly through the choked pin-hole.
2) When the inflator cools at night, the decreasing pressure in the inflator will draw humid air in through the pin-hole, and at least some of this moisture will be absorbed by the propellant.
3) This ingest/egest procedure repeats every day.

If enough moisture is absorbed, the propellant (gas generant) will weaken and crumble. The moisture safety limit is known from accelerated aging studies. The NHTSA requirement on gas-bag inflators was that they must work for at least 15 years, so the total moisture absorption over this period must be less than the maximum allowable amount for propellant integrity.

My Solution: I was asked to calculate the amount of moisture ingested over a 15 year period through a pin-hole of known diameter. The worst case scenario was assumed, i.e. 100% humidity with a daily diurnal temperature cycle from hot/humid New Orleans. I created a volume-filling code **GASBAG** that solved only the mass conservation equation for pressure history in the constant-volume inflator; the temperature history was specified by the imposed diurnal temperature cycle. The code was time-marched for a 14-day simulation, and validated by comparing its predictions for the special case of a step-function temperature cycle for which I had derived an approximate closed-form solution. Then the calculation was repeated using the real temperature cycle. The resulting histories of pressure and moisture ingestion exhibited a repeatable periodic behavior after a several-day start-up. The moisture ingestion rate became 0.0082 mg/day (i.e. 0.045 gm in 15 years). Since the flow through the pin-hole was choked, the solution was linear with hole area; the ingestion of 0.45 gm of water for a pin-hole with even ten times the area was still acceptable (Ref 23).

14-Day Diurnal Cycle of Moisture Ingestion in New Orleans

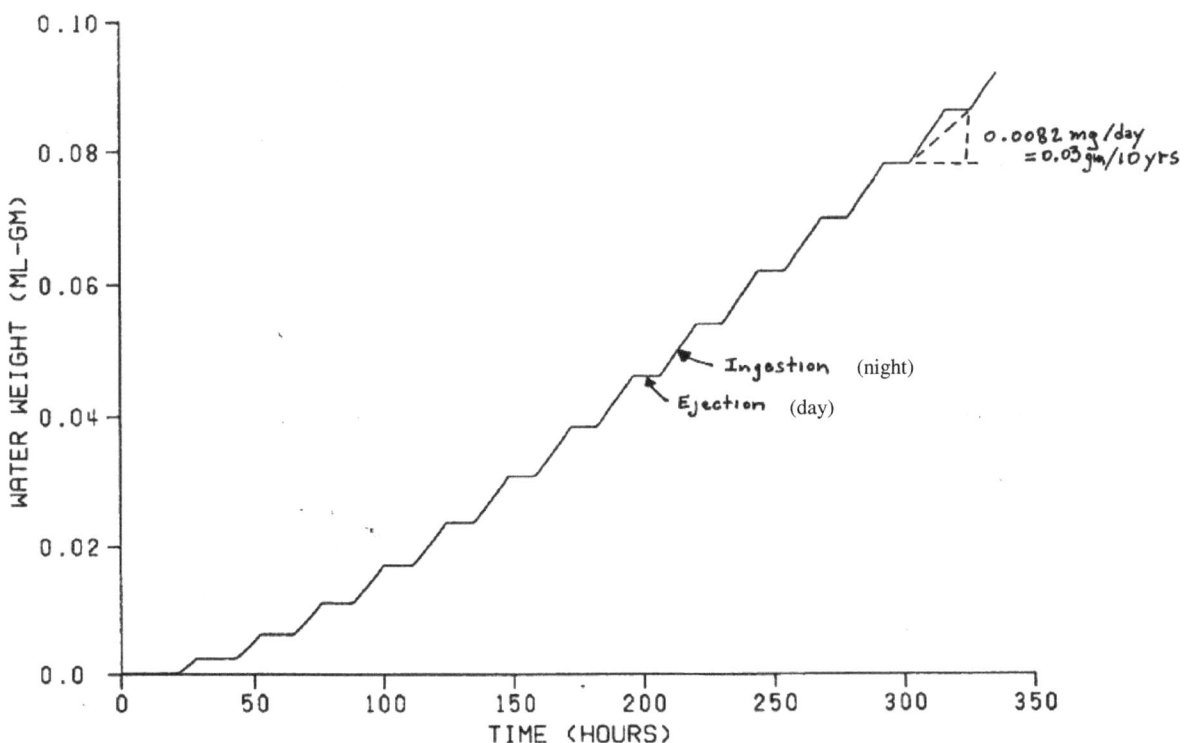

Recently (2015), there has been a spate of deaths caused by explosions of **Takata inflators**, especially in hot humid climates. Based on the study described above, I would not be surprised if they had a leak that was large enough to ingest an excessive amount of moisture. I contacted NHTSA to suggest this and to provide them with my model, but they never returned my call nor my email to their webmaster.

Vehicle Staging Analysis Using Volume-Filling Code STAGING

The Problem: Another application of the volume-filling and ignition methodology is the calculation of vehicle staging. Ignition of the upper stage pressurizes the interstage cavity (void volume between the motors in a multi-stage vehicle) and pushes the lower stage away. The equations for conservation of mass and energy within the cavity volume V determine the pressure history $p_{cav}(t)$ and temperature history $T_{cav}(t)$ of the gas in the cavity. The mass flow rate $\dot{m}_{in}$ and temperature T_{in} of the gas entering the cavity is generated by the <u>ignition of the upper stage</u>, and the mass flow rate $\dot{m}_{out}$ and temperature $T_{out}=T_{cav}$ is created by outflow through the skirt gap as the stages separate. The history of the skirt gap h(t) is determined by the force of the pressure p_{cav} pushing upward on the lower dome of the upper stage and downward on the upper dome of the lower stage, as well as the increasing thrust from the upper stage motor and the tailoff thrust of the lower stage.

Volume-Filling:

$$\frac{dp_{cav}}{dt} = \frac{RT_{cav}}{V}\left[\dot{m}_{in}-\dot{m}_{out}\right] + \frac{p_{cav}}{T_{cav}}\frac{dT_{cav}}{dt} - \frac{p_{cav}}{V}\frac{dV}{dt} \qquad \text{where } R = \frac{R_u}{\mu} \quad (7)$$

$$\frac{dT_{cav}}{dt} = \frac{RT_{cav}}{pV}\left[\gamma(\dot{m}_{in}T_{in}-\dot{m}_{out}T_{cav}) - (\dot{m}_{in}-\dot{m}_{out})T_{cav}\right] - \frac{RT_{cav}}{VC_v}\frac{dV}{dt} \qquad \text{where } \gamma = \frac{C_p}{C_v} \quad (8)$$

History of Separation Gap h(t):

$$\frac{d\delta}{dt} = \frac{F_{upp}+P_{upp}-D_{upp}}{m_{upp}} - \frac{F_{low}-P_{low}-D_{low}}{m_{low}} \quad \rightarrow \quad \delta = \frac{dh}{dt} \quad \rightarrow \quad h(t) \text{ and } \frac{dV}{dt} = \pi r_{cav}^2 \delta \quad (9)$$

where F is stage thrust, D is stage drag, m is stage mass, P is dome pressure force.

My Solution: I automated the solution of these equations as computer program **STAGING** (Ref 24) for use at Thiokol during the Peacekeeper program. However, I resurrected the code later at TRW for use in modeling of Minuteman subscale and full-scale staging events. For example, a comparison of the pressure history predicted by STAGING to that from the CFD code SALE3D (see page 58) and to measurement on the upper dome during a subscale test at CALSPAN is shown below. Also shown is a comparison of the predictions of STAGING and SALE3D for a full-scale staging event.

Although this lumped-parameter model does not capture the variation in pressure and temperature throughout the cavity, nor the pressure waves during the very-transient staging event (note the shock oscillation shown below predicted by the multi-dimensional code SALE3D), it has provided great insight and verification of the staging event. This includes the effects of skirt overlap, skirt vents, nozzle start-up shock ejection and re-ingestion, lower-stage tailoff rate, and upper stage ignition rate. Its simplicity and clarity drew praise from the USAF during a presentation made by my supervisor Steve Kovacic.

Comparison of Predicted Interstage Pressure Histories

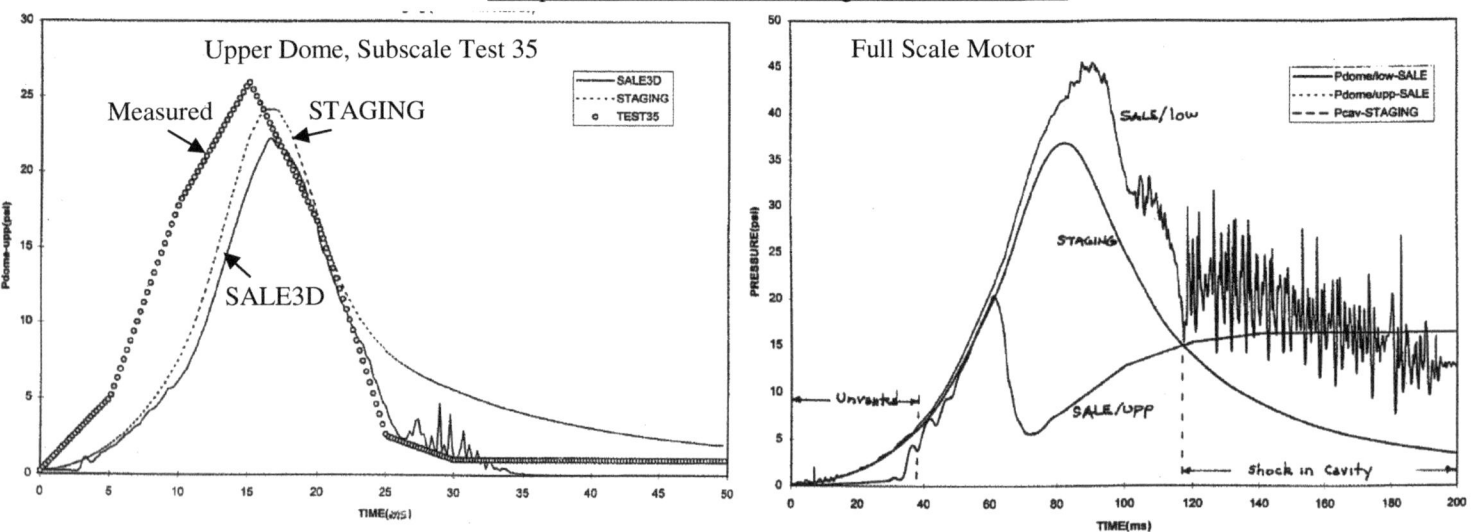

My colleagues wondered whether the high-frequency pressure oscillation predicted on the lower dome by SALE3D was a numerical anomaly or exposed a real phenomenon. I subsequently showed that it was a real phenomenon, caused after the startup shock from the upper stage pops into the interstage cavity and approaches the tip of the lower stage skirt (see figure on page 58); at this time, very small perturbations in the cavity pressure cause large changes in the standoff distance of the shock above the lower dome, which causes a large change in the pressure on the lower dome. I verified this hypothesis by verifying that the oscillation period was equal to the round-trip time of pressure waves reflecting between upper and lower domes. In addition, similar oscillations were predicted by other CFD codes FASTRAN and CFD++ (Chapter VIII in Ref 1).

(4D) <u>Motor Ignition and Its Impact on Shuttle Launches</u>

<u>Surprise Exposure of Igniter Shock and its Reflection Using Hoop Strain-Gage Data</u>

<u>**The Problem**</u>: High pressure inside a motor expands the case, which induces strain in the case wall. Ignition pressure histories <u>inside</u> statically-fired Shuttle boosters, deduced from hoop strain-gages wrapped around the <u>outside</u> of the booster cases at multiple axial locations (see page 33), at first appeared erratic (see figure below), and were initially ignored as unreliable.

<u>**My Solution**</u>: After some thought, I was able to explain this behavior: (1) When pyrogen igniters fire at the motor headend, they eject mass aftward so fast that they act like the driver in a shock tube. (2) Consequently, a shock wave propagates down the bore, which partially reflects from the nozzle inlet and aft dome. Despite the damping effect of the case and propellant masses, hoop strain measured at multiple axial locations revealed this propagation.

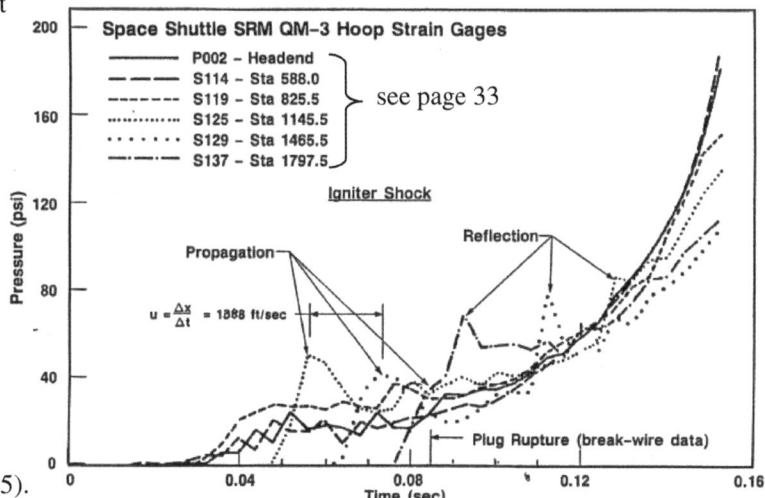

By superposing the known time at which the break-wire across the nozzle throat weather seal (throat plug) ruptured due to the impact of the igniter shock, the <u>wave reflection</u> became obvious. The "spikes" in pressure exposed the passage of the shock past each strain gage. As further validation, the measured speed of the shock (distance Δx between gages, divided by the difference in their times Δt at peak pressure) was shown to agree well with predictions from a 1D CFD ignition code. This was a break-through in explaining multiple chamber phenomena: knees in the head-end pressure trace (see page 65), rupture of the weather seal (plug), as well as shock expulsion from the nozzle (Refs 1, 25).

In addition, after rupture of the plug at 85 ms, part of the shock propagates down the now-open nozzle and exits the nozzle at about 90 ms. This fact helped explain the ignition overpressure (IOP) in the first Shuttle launch STS-1 described below.

<u>Ignition Overpressure (IOP) During First Shuttle Launch in 1981: Near Disaster</u>

<u>**The Problem**</u>: During the first launch of the Space Shuttle (STS-1) in 1981, <u>the ailerons on the orbiter wings were nearly knocked off</u> even before lift-off by a strong pressure wave created by the ignition of the solid-propellant boosters. A strong pressure wave had been expected, so a water spray system had been installed in the <u>underground</u> exhaust trench to scatter and damp the wave. NASA stopped the Shuttle program until a cause of the excessive wave and a fix were identified.

<u>**My Solution**</u>: NASA had always believed the igniter to be unnecessarily strong. Consequently, at first they blamed the igniter. However, using the shock behavior shown above, I showed that the igniter shock had exited the nozzle much earlier, whereas the dangerous 3.0 psi over-pressure wave was instead caused by ignition of the booster main propellant (see figure below). Secondly, I showed that the wave reflection occurred <u>inside the exhaust holes in the Mobile Launch Platform</u> (MLP), rather than from the underground exhaust trench (Ref 26):

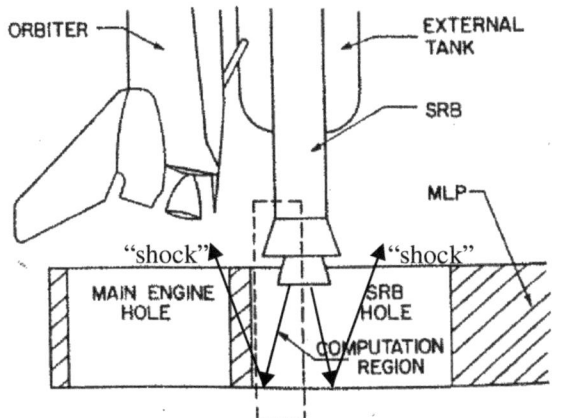

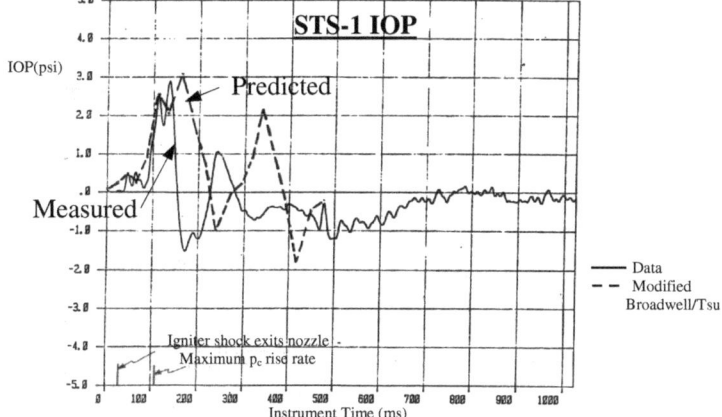

Since it was too late to redesign the boosters to soften the pressure wave, I recommended that the only fix would be to move the water spray (meant to dampen the pressure wave) from the exhaust trench <u>up into the MLP holes</u>, and triple its flow rate.

> My analysis was sent by NASA to many organizations for review, but none found fault with it. This is Example #2 of "**You never know who will be reading your reports, so write them well!**" The proposed fixes were first validated in subscale testing, then implemented on the MLP. The Shuttle launch schedule resumed after a year's delay. The mixing of the water spray with the hot rocket exhaust caused the vast steam cloud seen during every subsequent Shuttle launch.

(4E) O-Rings and the Challenger Accident

O-Ring Erosion in Shuttle Boosters

The Problem: Prior to the ill-fated Challenger flight in January 1986, periodic erosion of some of the O-rings in both case and nozzle joints of the Shuttle booster rockets had been observed in both static tests and flights (18 O-rings out of 232 on the 58 boosters previously flown). The cause was known, but not modeled as of April 1985.

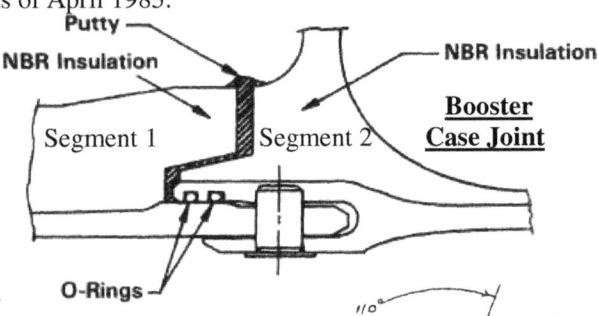

The joint gaps between the segments of the solid boosters were filled with zinc-chromate putty in an attempt to keep the hot gas/droplet flow away from the primary and secondary O-rings that sealed the joints. Unfortunately, during assembly of the segments, air trapped in the joint punched narrow "blow holes" through the putty as the segments were pressed together. Some of these blow holes would penetrate all the way to the chamber; in these cases, during motor ignition, hot gas/particle ignition products would flow through the blow holes from the chamber to the joint and impinge as a jet on the primary O-rings, causing erosion.

My Solution: Hardware engineer Roger Boisjoly at Thiokol asked me to attempt to explain and quantify the erosion phenomenon for the **case-joint** O-rings. I was able to model the joint pressurization process using simple volume-filling equations, and to model erosion depth using Shadlesky's model of impingement heating. The model showed that the degree of impingement erosion increased as the width of the blow hole decreased, i.e. as the jet concentrated. However, this impingement lasted only for the 600 msec until the joint was pressurized to the same pressure as the chamber. As long as the joint remained sealed, a limited amount of heat would enter the joint during this 600 msec, and the surrounding metal would heat less than 100°F.

Agreement with measurement was excellent as long as the width of the putty blow hole was known (as it was upon joint disassembly <u>after</u> static test or post-flight recovery). However, the width of the blow hole was <u>random</u>, and therefore <u>unpredictable</u> for the next booster due to variations in assembly. I presented these results to the engineers and managers at the NASA Marshall Space Flight Center (MSFC). My reports were well received by both Thiokol and NASA management.

STS-41C O-Ring Failure: The Game Changer

The Problem: NASA and Thiokol believed it was safe to fly the Space Transportation System (STS) despite the O-ring erosion experienced on numerous flights because the erosion seemed to leave sufficient safety margin, and the joints also contained secondary (backup) O-rings which had never experienced erosion. Previous to July 1985, no primary O-ring had ever burned completely through. However, in that month, upon disassembly in Layton Utah of a nozzle from flight STS-41C, it was discovered that the primary **nozzle** O-ring had indeed completely burned through. That was a game changer.

My Solution: It was a Tuesday morning. I had stayed home because I was scheduled to fly that afternoon to MSFC to present my latest modeling results for erosion of **case** O-rings. Normally I would have been playing ice hockey in Bountiful Utah that evening, as I did every Tuesday, so I had told my teammate Bill Green with whom I normally car-pooled to hockey that I would not be playing that week. All that changed when the phone rang in the morning telling me not to fly out that afternoon, but instead to go to the Layton facility to inspect a failed nozzle primary O-ring and its joint. I called Bill and told him that I would be able after all to go to hockey that evening, but I would have to stop for a while in Layton, which was on the way. So Bill waited in the car while I examined the O-ring and joint. I was told that I would have to model **nozzle** O-ring erosion by Thursday because there was a Flight Readiness Review (FRR) for the next scheduled flight on Saturday. Starting after I returned from ice hockey, and going almost continuously until the FRR on Thursday afternoon, I developed a model of nozzle O-ring impingement erosion. I concluded that the nozzle secondary (backup) O-ring was "safe" even when the primary O-ring burns through because the secondary O-ring was around a 90°-corner in the joint; thus the hot jet passing through any failed primary O-ring would not impinge directly on the secondary O-ring, but would have spread circumferentially and weakened greatly. I hurriedly generated the viewgraphs to explain the model and the conclusions, and sent them to Al McDonald, Thiokol director of the Space Shuttle SRM Project.

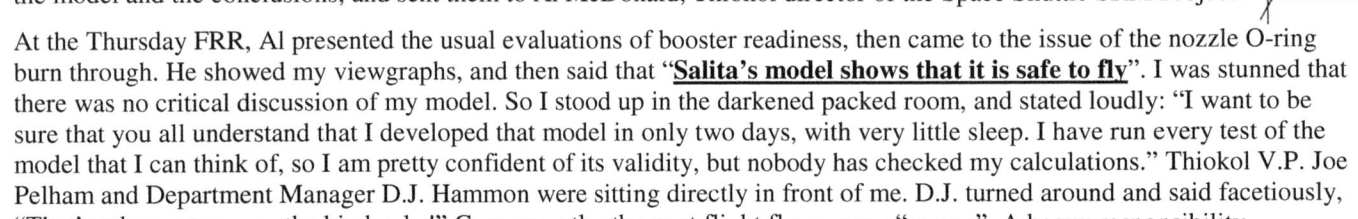

At the Thursday FRR, Al presented the usual evaluations of booster readiness, then came to the issue of the nozzle O-ring burn through. He showed my viewgraphs, and then said that "**Salita's model shows that it is safe to fly**". I was stunned that there was no critical discussion of my model. So I stood up in the darkened packed room, and stated loudly: "I want to be sure that you all understand that I developed that model in only two days, with very little sleep. I have run every test of the model that I can think of, so I am pretty confident of its validity, but nobody has checked my calculations." Thiokol V.P. Joe Pelham and Department Manager D.J. Hammon were sitting directly in front of me. D.J. turned around and said facetiously, "That's why we pay you the big bucks!" Consequently, the next flight flew on my "say-so". A heavy responsibility.

Subsequently, Al presented a summary of the O-ring erosion problem, including my viewgraphs, to NASA Headquarters on August 19 1985, and pleaded with them that STS-41C showed that we needed to redesign the joints to avoid any O-ring erosion at all. This convinced NASA to allocate funds to support a Joint Redesign Team ("O-Ring Seal Task Force"). I was one of the six members on that team. The first step was to remove the putty that formed the "blow holes" that caused the jet-impingement erosion. We studied replacing the putty with staggered NBR insulation strips, or porous rope that would remove the heat from the gas entering the joint, yet allow the O-rings to pressurize and seat. Unfortunately, before that was accomplished, erosion of an excessively-cold O-ring at the launch of Challenger flight STS-51L in January 1986 allowed hot gas to leak past both the primary and secondary O-rings in the aft <u>case joint</u> of one booster, jet onto and melt the aft strut holding it to the External Tank (ET) containing tons of liquid hydrogen and oxygen, thereby allowing the booster nose to smash into the ET and causing an explosion (more precisely, "deflagration").

My detailed and widely distributed reports (covers shown below) proved that O-ring erosion had been taken seriously by Thiokol and NASA before Challenger, thus protecting Thiokol from numerous subsequent law suits:

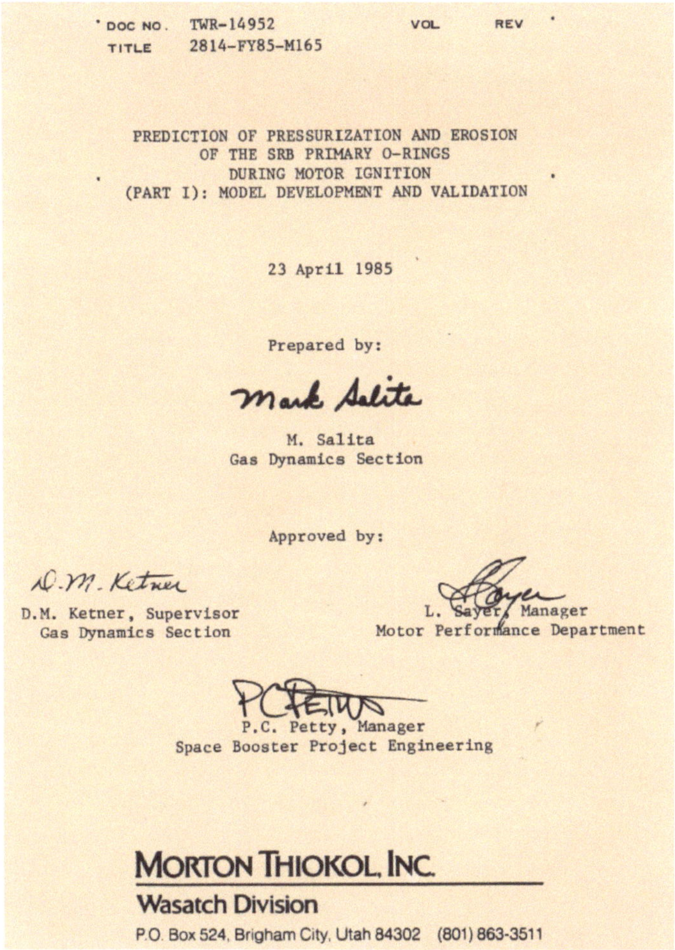

DOC NO. TWR-14952 VOL REV	DOC NO. TWR-15186 VOL REV
TITLE 2814-FY85-M165	TITLE 2814-FY86-M015

PREDICTION OF PRESSURIZATION AND EROSION
OF THE SRB PRIMARY O-RINGS
DURING MOTOR IGNITION
(PART I): MODEL DEVELOPMENT AND VALIDATION

23 April 1985

Prepared by:

Mark Salita

M. Salita
Gas Dynamics Section

Approved by:

D.M. Ketner   *L. Sayer*
D.M. Ketner, Supervisor   L. Sayer, Manager
Gas Dynamics Section   Motor Performance Department

P.C. Petty
P.C. Petty, Manager
Space Booster Project Engineering

MORTON THIOKOL, INC.
Wasatch Division
P.O. Box 524, Brigham City, Utah 84302   (801) 863-3511

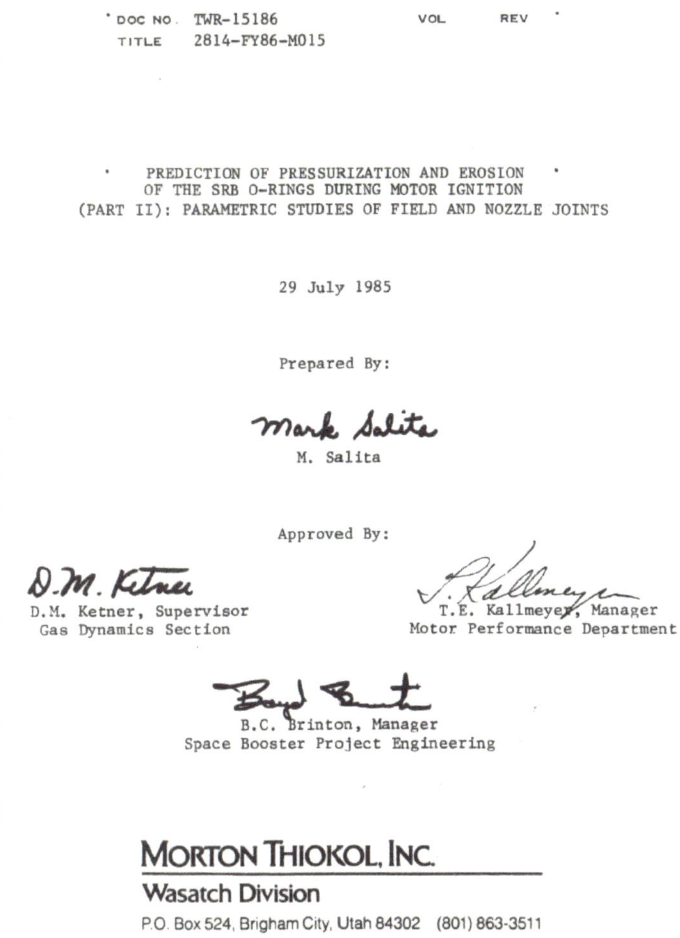

PREDICTION OF PRESSURIZATION AND EROSION
OF THE SRB O-RINGS DURING MOTOR IGNITION
(PART II): PARAMETRIC STUDIES OF FIELD AND NOZZLE JOINTS

29 July 1985

Prepared By:

Mark Salita

M. Salita

Approved By:

D.M. Ketner   *T.E. Kallmeyer*
D.M. Ketner, Supervisor   T.E. Kallmeyer, Manager
Gas Dynamics Section   Motor Performance Department

B.C. Brinton
B.C. Brinton, Manager
Space Booster Project Engineering

MORTON THIOKOL, INC.
Wasatch Division
P.O. Box 524, Brigham City, Utah 84302   (801) 863-3511

A detailed discussion of the O-ring problem and Thiokol's interaction with NASA has been provided by Al McDonald in his book <u>Truth, Lies, and O-Rings</u>. That is the story from his perspective, and is very detailed and accurate. Although he did reference my modeling work, he didn't document the degree to which it played a roll in both Thiokol's and NASA's decision process. For example, many of the viewgraphs that he presented to NASA Headquarters were mine.

<u>Strange coincidence</u>: Engineers named McDonald played an interesting part in my life. Al McDonald was my first tennis partner when I moved to Utah, and his daughter was my daughter's first baby sitter there. Previously, the well-know Harry McDonald (UTC group head, founder of SRA, and one-time Director of NASA Ames) had helped get me the job at Pratt and Whitney by recommending me to Bob Dring (see page 12), and Harry's wife had been my daughter's first pediatrician in Connecticut.

<u>Cute Story</u>: I was sitting at the dining room table one evening debugging my O-ring deformation code.
To simplify the hand calculations, I was approximating the O-ring cross-section using only four elements.
My 10 year-old daughter, who was sitting across the table, asked "Daddy, why are you drawing a kite?"
I explained, and the debug was successful. Results for 72 finite elements are shown on page 30.

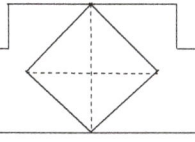

Challenger Flight Approval, Accident, and Subsequent Investigation: My Story

The afternoon before the launch of Challenger, I was told to report to the main conference room to discuss the effect of the cold air temperature predicted for the time of launch the next morning. All the appropriate Thiokol engineers and managers were at the meeting. NASA managers at MSFC and Al McDonald at Cape Kennedy were on the speaker phones. The resulting arguments between Thiokol and NASA have been discussed in detail in many other places. However, I will add my two cents here.

Everyone understood that cold O-rings were less resilient than warm ones. The previous coldest launch was at 53°F, and significant soot deposits had subsequently been observed between the primary and secondary O-rings, suggesting incomplete sealing of the primary O-rings at this "low" temperature. I was seated opposite the two hardware specialists (Roger Boisjoly and Arnie Thompson) who pleaded to delay the launch based on the fact that the predicted temperature at launch (26°F) would be much lower than the previous coldest, and the soot was an omen. As the world knows, NASA convinced Thiokol managers (except Al McDonald who was also afraid of the dangerous ice formations on the launch tower in addition to the cold O-ring concerns) to give a green light for the boosters despite these concerns. After Challenger blew up the next day, Boisjoly suffered from PTSD, feeling that the accident was his fault for not fighting hard enough to delay the launch. There was no way he and Arnie could have altered the decision and subsequent disaster. Ironically though, the disaster was not due simply to the cold air temperature. My model of O-ring activation developed after the accident showed that there is always a short period of O-ring blowby (which would create soot due to burnt putty and grease), but the primary O-rings would seal successfully even at 26°F. The failure of both O-rings in only one of the six case joints on the pair of boosters was due to the unique stack-up of four other detrimental factors, unknown at the flight review that fateful night (discussed in more detail on pages 31-32).

Initially, it was unknown what caused the accident. Rockwell was at first sure that it was due to failure of the Space Shuttle Main Engine (SSME) that augmented the SRMs (as told by the Rockwell CEO to Boisjoly on the way to MSFC). The SSME operated at a very severe cycle environment, and had experienced many test failures during development. However, review of one of the 70mm videos taken during the flight revealed a jet of hot gas emanating from the booster and impinging on the External Tank (ET). When Thiokol CEO Jerry Mason was notified of this, he called me and Phil Shadlesky into his office to tell us that we had to try to understand this new development. At the same time, President Reagan created a Presidential Commission headed by former Secretary of State William Rogers to investigate the accident. Several days later, the media and the Commission got wind of the fact that there had been O-ring erosion all along, yet the flight schedule had continued.

One night shortly thereafter, a telephone operator interrupted my wife's conversation with her aunt, to say that she was asked to put an important call through to me. It was Al McDonald, and he told me to pack my bags because he would be picking me up in an hour to take us to the local airport, from which we would be flying on the company jet to Washington DC that night in order to attend a suddenly-called first meeting of the Presidential Commission the next morning. I did, and we flew to Dulles Airport in only two and a half hours due to high tail winds (coincidently, a 737 flight between Salt Lake City and Washington DC that same night set a record for the fastest ever such flight).

A limo was waiting on the tarmac when we arrived at Dulles at 3AM, and took us to a hotel for a few hours sleep before the meeting. Besides Al and myself, Thiokol CEO Jerry Mason and Vice President of Manufacturing Cal Wiggins were on the flight. Jerry told us that only Al, myself, and soon-to-arrive head of the redesign team, Don Ketner, would attend the Commission meeting, and that we were not to say anything unless asked a question.

The next morning we attended the Commission meeting at the State Department. Besides us, attendees included NASA managers Larry Mulloy (Program Manager for the SRB at NASA's Marshall Space Flight Center) and Larry Wear (NASA SRM Manager), and Commission members Rogers, astronauts Neil Armstrong and Sally Ride, Nobel Laureate Richard Feynman, General Donald Kutyna, USAF Chief Scientist Gene Covert, and late-arriving test pilot Chuck Yeager.

Larry Mulloy presented NASA's summary of the O-ring problem, but he was so confusing that the Commission members accused him of attempting a cover-up. This bad situation continued on and on. Meanwhile, I knew that I could explain the issue more clearly, but was frustrated by the requirement to keep quiet unless called upon by NASA. I was holding onto the bottom of my seat until my fingers must have been turning blue, wanting to jump up and say "let me explain the problem". It turned out that Al McDonald, who was sitting next to me, had similarly been holding onto his seat. Finally, he spoke up and told the Commission that Thiokol had recommended the night before the launch that the launch should be delayed due to the low temperature, but that NASA had cajoled the Thiokol managers other than himself into approving the launch. Rogers then asked NASA to prepare a presentation explaining the decision to launch that would be televised in front of the media the next morning. And he wanted a list of all attendees that had been at the meeting when flight approval was given.

That night we all retired to NASA Headquarters for the NASA managers to put together the requested presentation. Also joining us was the feared head of NASA Marshall Space Flight Center (MSFC), Dr. William Lucas. I had previously made several presentations to him and the "Lucas Board" at MSFC describing my model of O-ring impingement erosion. But this night Lucas seemed to be catatonic, due to the stress of the accident, not knowing what to do. Despite the order that I not participate unless asked by NASA, in my frustration, I put together a set of viewgraphs summarizing the history of O-ring erosion, the modeling, and its implications, and handed them to Lucas and said "this is what you need to present". My boss Don Ketner wasn't too happy that I violated my instructions.

The next morning, Larry Wear presented viewgraphs, many of which were polished versions of those that I had recommended the night before. The Commission members said "now that's what we wanted to see!" I smiled at Ketner. Soon after, there was a media break, and the Commission members, NASA representatives, and Thiokol engineers moved into a refreshment area. I saw Richard Feynman standing there with a Styrofoam cup containing ice water and a section of O-ring squeezed by a C-clamp. I went up to him and said "I see you're going to demonstrate the reduced resiliency of a cold O-ring." He responded "What do you think of this?" I replied, "It's informative but it's not accurate because there is no high gas pressure on the unclamped side." He left to get a donut. Upon a reconvening of the media event, Feynman showed the world the cold O-ring in the C-clamp, a cute moment. Within the next month I would develop a computer program that would quantify the O-ring behavior as a function of temperature (summarized on pages 94-95).

First however, I had to spend several weeks at MSFC to support the accident investigation and the joint redesign effort. As part of the investigation, I was asked to supply to the Presidential Commission my reports describing my previous modeling of the O-ring erosion history. Several days later, I was asked to meet with members of the Commission so they could ask me questions about my models. At that meeting, Richard Feynman sat on my right, Gene Covert on my left, Neil Armstrong and Sally Ride sat across the table, and General Kutyna sat at the end of the table. I answered their questions about why I had made various assumptions, etc. Those reports (see covers on page 27) from the previous year made it clear that there had been no cover-up or ignoring of the erosion problem. Of course, when I wrote those reports, I had no idea that they would be read by such a prestigious group. I was glad that I had done a careful job. **You never know who will be reading your reports** (Example #3)!

The next day Shadlesky and I were walking down a hallway at MSFC when a technician came out of a door. He recognized us, and said "come see some photos that I have in my lab". He then showed us pictures of Challenger on the launch pad taken the night before the launch. **This chance encounter and the pictures would be the key to understanding the real cause of the Challenger accident.** They revealed a **fog-like stream** passing over the ET and diving down onto the bottom of one nearby attached booster. The "fog", providing unintended "flow visualization", was caused by frozen droplets of water being blown by a stiff wind from the eye-wash fountain on the Service Tower, left running all night so that it wouldn't freeze. The flow over the ET was "super-cooled" by its cryogenic liquid propellant, sank due to negative buoyancy, and impinged on the booster aft joint and cooled it to a temperature of only 8°F, as recorded by the IR-camera used by the Ice Team the night before the flight, but not reported to us. I'm still kicking myself for not getting copies of those photos then; subsequent attempts to find them in NASA archives have failed.

I now knew that the joint with the failed O-ring had been supercooled without the knowledge of the engineers and managers in charge of deciding whether to launch Challenger. When I returned to Utah, I recommended that our CFD experts model the launch pad thermal environment the night before the launch, and that I would create a model of pressurization and activation of an O-ring in a groove as a function of its temperature. No such model had ever been derived. We needed quick results, so I asked structural analyst Dr. Ron Webster for a simplified model of material deformation (since I was only a fluid dynamicist). He gave me a linear small-deformation model with the warning that it would be formally invalid for the large deformations encountered by the O-rings. However, I needed any model quickly that I could couple to my joint pressurization model. Luckily, I was able to code the model (ORINGDEF) and show that the predictions matched a huge range of data obtained during the accident investigation and the joint redesign. This surprising success was due to the dominant constraints provided on the material deformation geometrically by the groove walls surrounding the O-rings (see next page), rather than by the non-linear viscoelastic O-ring material properties.

The story of how I constructed this solution is explained in greater detail on pages 94-95. I presented the results at an AIAA conference and in the Journal of Propulsion (Ref 27), which was read around the world. The computer code that I wrote has since been used by other U.S. organizations including UTRC, TRW, Raytheon, and the nuclear waste containment industry.

A revealing incident happened to my daughter Karen on a flight to her employer's headquarters in Dallas. She and her airplane seat neighbor exchanged names, and the man asked if she was related to Mark Salita. She replied that he is her dad. The man was a nuclear physicist and probably read my O-ring journal article for his nuclear containment work.

The O-ring behavior predicted by ORINGDEF during the booster ignition transient (600 ms) is shown on the next page for O-ring temperatures of 75°F and 25°F in the expanding joint gap created by booster pressurization. Note that the temporary blow-by is what generated the soot on the previously-coldest launch (and indeed on many other boosters at ignition), as well as the puff of black smoke at Challenger launch. Even at 25°F the joint is predicted to seal. However, the joint was apparently at 8°F, and its O-ring failed to seal initially and suffered damage. Subsequent subscale testing showed that O-ring blow-by damage can be sealed like a dam by the Al_2O_3 droplets in the flow. Indeed, the motor chamber pressure was totally normal for the first 58 sec of flight. So I postulated that the joint was sealed until Challenger flew through the severe wind-shear at 58 sec, shaking the booster, thereby rupturing the "dam" (see the following article from Aerospace America (Ref 3)).

29

ORINGDEF simulation of activation of O-ring in the expanding aft field joint of a Shuttle booster at two temperatures: (a) no blowby predicted at 75°F (resilient), and (b) temporary blowby predicted at 25°F (non-resilient).

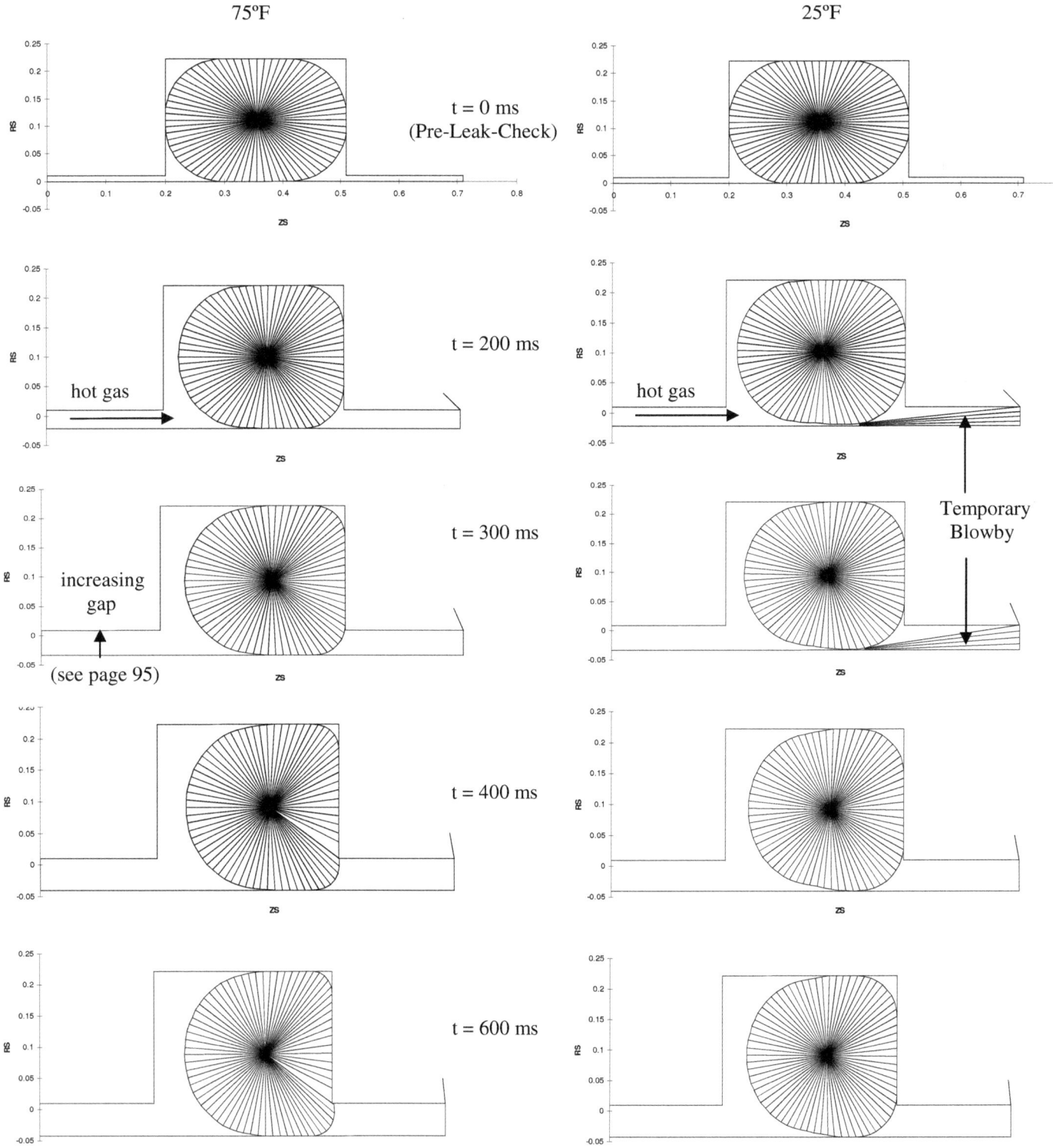

75°F   25°F

t = 0 ms (Pre-Leak-Check)

t = 200 ms

hot gas

t = 300 ms

increasing gap

(see page 95)

Temporary Blowby

t = 400 ms

t = 600 ms

The above animation was shown on national TV shortly thereafter, and was the first engineering (non-cartoon) simulation.

There are practical benefits to understanding how O-rings work. One day I noticed that the kitchen faucet was dripping and the water pressure was low. I suspected that the low pressure was preventing the faucet O-rings from seating properly. So I told my wife to fill pots with water because I thought the city water might be turned off shortly. Indeed that happened.

Several years later, a plumber replaced our bathtub inflow valve. That evening I noticed that the tub faucet was dripping. My wife immediately suggested that the plumber had not installed the O-rings properly. When the plumber returned, he found that he had indeed installed them incorrectly.

Wind Shear During Vehicle Ascent:
The Cause of Both Challenger and Columbia Accidents?

The following accident analysis by Mark Salita was published in Aerospace America in 2004 (Ref 3).

The Columbia Accident Investigation Board (CAIB) has released its report on the disastrous reentry of Shuttle flight STS-107. This report blames NASA management practices to be as much a cause of the accident as the foam that struck the left wing 81 sec into flight. These practices included allowing flight with known flaws, blocking the flow of critical information up the hierarchy, and inadequate safety monitoring. Chapter 8 of the report documents the similarities of the events and management decisions that led to both the Challenger and Columbia accidents. In particular, the report noted that the success of Shuttle missions in spite of persistent O-ring erosion prior to 1986 and wing impact of foam debris throughout the program led to the acceptance of these flaws.

Unfortunately, the report failed to mention an important similarity between the two accidents: Columbia and Challenger flew through two of the three worst wind shears of all Shuttle flights (the third being STS-90), at nearly identical times (57-58 sec) close to that of maximum vehicle dynamic loading ("max Q"). The CAIB did report that the aerodynamic loads induced by the wind shear during Columbia's ascent were only 70% of the design certification limit for the Orbiter wings and ET forward attachment. However, no models were available at the time of the CAIB report to assess the effect of the wind shear on the foam/ET bondline, although one is currently under development. In addition, no account has been taken of the possibility that a coupling of a series of detrimental factors, the last (or even first) one being the wind shear might have caused the accident. Even if these secondary factors could not have caused the accident in the absence of wind shear, and the wind shear could not have caused the accident in the absence of the secondary factors, it is possible that the simultaneous occurrence of both could have been fatal. Indeed, this is exactly what happened to Challenger.

Analysis of the Challenger flight has shown that the accident was not the result of a single event, but of a sequence of unfortunate events that was never anticipated, the last of which was the fatal wind shear. Although the individual events were discovered during the Challenger accident investigation, few people are aware of the significance of their sequence and interaction. Consequently, it is important to review the critical combination of factors that caused the Challenger accident in order to understand better those that might have caused the Columbia accident, and perhaps more importantly, might conceivably cause future Shuttle failures.

The External Tank (ET) on Space Transportation System (STS) 51L (Challenger) exploded 76 seconds after its launch. The cause of this explosion has been blamed on the failure of both the primary and secondary cold O-rings to seal the aft of three case-joints of one of the solid propellant boosters. Most people believe that this failure was due simply to the cold 28°F air temperature at launch. This is not true. The cold air temperature did reduce the resiliency of the O-rings, but not enough to cause the O-rings to fail to seal their joints at launch; indeed, seven of the eight case and nozzle joints on the boosters sealed without incident. The accident only occurred because of an unfortunate stack-up of five detrimental events, the first of which was air temperature. Had any of the other four not occurred, there would not have been an accident.

The second detrimental event was the presence of a steady wind at the launch pad for 48 hours preceding the launch of 51L. This wind flowed over the ET and was super-cooled in two ways: (1) by the very cold ET surface that contains cryogenic propellant (liquid hydrogen at minus 423°F and liquid oxygen at minus 300°F) that generates a layer of ice even in summer, and (2) by mixing the air with the liquid oxygen vapors that are continually venting at minus 120°F due to its boiling inside the ET during fueling. The resulting super-cooled air then dove toward the ground behind the ET due to its negative buoyancy (exactly the opposite behavior to the positive buoyancy of a hot-air balloon, and familiar to anyone who stands with bare feet in front of an open refrigerator). This diving flow was made visible by the entrainment and crystalization of the water spray from the eye-wash fountain on the Shuttle service tower, and was captured in photographs developed only after the accident. The presence of this super-cooled air was further manifest by the half-inch of ice in the water hammocks below the boosters, despite the addition of antifreeze that reduced its freeze temperature to 9°F. If there had been no wind, the third event described below could not have occurred and the Challenger accident would not have happened.

The third detrimental event was the direction of the wind. It was blowing in such an unusual and unfortunate direction (from west/northwest) that the diving cold air was impinging on one of the solid boosters near its aft joint and directly facing the ET. Furthermore, this direction persisted for the full 48 hours, whereas wind direction at the launch site normally varies randomly. Thus, the O-rings in this aft-joint facing the ET were far colder and less resilient than in any of the other seven. Numerous thermal analyses were subsequently conducted that verified that the temperature of the O-rings in the aft joint was only 2-10°F at the time of launch! This allowed temporary blow-by and damage to these O-rings to occur at motor ignition. Even so, these O-rings did seal after the blow-by, and maintained their seal for the first half of booster operation. If the wind had been blowing in almost any other direction during the hours before launch, the aft joint would not have been super-cooled or would have been super-cooled at a location not directly facing the ET, and the Challenger accident would not have happened.

The fourth detrimental event was a human error of judgment. An Ice Team measures the thickness of ice on the ET before every launch using an IR camera. A temperature of 8°F was by chance recorded at the booster aft joint on 51L by the Ice Team, who reported it to their supervisor. Unfortunately, this warning was not passed up the chain of command because it was the Ice Team's responsibility only to estimate the thickness of ice on the ET. If the NASA/Thiokol launch team had been told of the 8°F measurement on the booster, the launch would certainly have been aborted and the Challenger accident would not have happened.

Even with these four detrimental events, the damaged O-rings in the aft joint still provided a complete seal for 58 seconds of flight. Evidence of this seal was provided by the nominal history of chamber pressure in the booster during this time. The puff of black smoke ejected from the aft joint at booster ignition showed severe but temporary blow-by of the O-rings. By itself, this smoke did not prove that there was O-ring damage. Indeed, many missiles generate clouds of black smoke at ignition due to temporary blow-by of their O-rings and burning of their grease coating, but subsequent examination of these O-rings after static testing typically shows no damage or even heat effect. The ability of even damaged O-rings to seal a booster joint has been demonstrated by the results of many test firings that have shown that O-rings, even with notches cut out of them, often do not leak due to plugging by aluminum oxide particles that form during combustion of the aluminized booster propellant.

The final contributing event occurred 58 seconds after launch when the Shuttle passed through the worst wind shear experienced by any STS flight as of that date. Even worse, this timing nearly coincided with the "max Q" condition of maximum vehicle vibration. This event flexed the booster so severely that it dislodged the aluminum oxide dam that had plugged the hole in the damaged O-ring during the first 58 seconds of flight. The pressure in the motor chamber only then began to drop from its nominal history (indicating the onset of a leak). The leak allowed 6000°F chamber gas to flow past the O-rings, to melt the joint and burn a hole through the motor case, to jet through this hole and burn through the ET and attach-strut. This allowed the forward end of the booster to swing free and rupture the LOX tank, whose oxidizer reacted catastrophically with the leaking liquid hydrogen fuel. In the absence of the wind shear, the damaged but plugged O-rings probably would have survived the 120 seconds of booster operation without leaking, and the Challenger accident would not have happened. Even with a leak starting at 58 seconds, but not pointing at the ET, the booster would probably have continued to function for its remaining 62 seconds of operation (although with reduced chamber pressure and thrust). Indeed, the leaking booster continued to fly on course even after the explosion of the ET, until the range safety officer destroyed it. The desired orbit would not have been achievable, but Challenger could have landed safely at its pre-programmed abort site.

In summary, the Challenger accident would not have occurred in the absence of the wind at the launch pad. Even with the wind, the accident would not have occurred if its direction had been different. Even with the wind in its worst direction, the flight would have been aborted to a safer time if the Ice Team had reported the 8°F temperature measured on the booster. Even with the launch that did occur, the flight probably would have proceeded normally in the absence of the wind shear.

It is important that the Columbia investigation consider similar interactions. For example, the wind shear might have been the last straw for a bondline weakened by substandard or misapplied foam, or by ET/Orbiter misalignment (seen at ET separation), or by the cooler than normal temperature of the January 16 launch (intriguingly within 12 days of the January 28 launch of Challenger). Even though water retention tests of the foam have shown that its density increases less than 2% due to immersion in water, does the frost observed earlier in the morning indicate that the foam around the shaded bipod ramp might have been denser than usual due to increased content of crystalized moisture or thicker ice?

Or the wind shear might have started the fatal sequence. For example, the CAIB noted that the wind shear induced a sloshing in the LOX tank that peaked at an amplitude at 75 sec that was higher than normal, and lasted longer. And what about the effect of the "Performance Enhancement" event that was activated at 80 sec to increase axial thrust by vectoring the SSME nozzles? Consequently, it seems possible that the severe flexing of the STS system by the wind shear could have loosened the foam directly, or loosened the foam indirectly by increasing the amplitude of the LOX slosh and the resulting vehicle vibration, which further enhanced by the nozzle vectoring at 80 sec might then have caused the fatal foam ejection at 81 sec.

If this coupling seems far fetched, remember the Challenger scenario, and ask: what unforeseen part of the STS system might fail the next time the vehicle passes through a severe wind shear? Since it is probably impossible to model or simulate experimentally these and other as-yet unidentified interactions, STS flight through severe wind shear must be avoided even at the expense of frustrating launch delays.

(4F) <u>Motor Pressure Surprises</u>

<u>Erroneous NASA Slot Pressure Prediction and My Fix</u>

<u>**The Problem**</u>: Several years into the Shuttle booster program, a NASA engineer applied a Boeing computer program to predict chamber pressure throughout the Shuttle booster. It predicted that the pressure in the propellant intersegment slots was higher than allowed. <u>NASA immediately halted production of the boosters</u> (see page 11 of Ref 2).

<u>**My Solution**</u>: I was asked to review his predictions. I showed that his program, which predicted <u>stagnation</u> pressure at the bottom of the slot, was not correct due to the highly-viscous slot flow; the real slot flow would have a much lower and allowable <u>static</u> pressure like through a boundary layer. NASA sent my analysis to many organizations for review, but none found fault with it. So <u>NASA gave the OK to restart the booster manufacturing</u>.

Another example (#4) of the axiom "**You never know where your report will end up, so write it as well as possible**" occurred when I called the Johnson Space Center (JSC) during this same time period to inquire about a different project. My phone call was passed on to an engineer C.P. Lee whose name I had never heard previously.

    I said "My name is Mark Salita, and you don't know me."

    He replied "Yes I do. I have a copy of your analysis of the slot pressure on my desk, and I was just reviewing it."

Schematic of Intersegment Slot
(Exaggerated Width)

vortex
stack

static pressure, not stagnation

<u>Viscoelastic Propellant Response Determined from Hoop Strain Gauges</u>

<u>**The Problem**</u>: For a number of years, NASA prodded Thiokol to explain the cause of the loop in the plot of hoop strain e_θ versus chamber pressure p (see figure below for example at strain gage S585) observed from girth (hoop) gauges wound around Shuttle booster cases during all static tests. Initially, it was believed that it was caused by a hysteresis effect.

<u>**My Solution**</u>: The plot of rocket chamber pressure p versus case hoop strain e_θ should be a straight line $p=Ke_\theta$ if the case and propellant are elastic (spring-like). However, rapid pressurization of the motor during the ignition transient causes the visco-elastic propellant to resist deformation, like a viscous damper in a car (the faster you squeeze it, the more it resists). With the experience of modeling the viscoelastic response of O-rings using a spring-dashpot analog (see pages 94-95), I modeled the dependence of chamber pressure p on hoop strain <u>and strain rate</u> of the propellant in a similar manner (Ref 25):

$$p = K[e_\theta + \tau \dot{e}_\theta] \qquad \tau = \alpha \dot{p} = \text{dashpot (resiliency) parameter} \qquad (\dot{\ }) = \text{rate } d/dt$$

where K is the stiffness (spring constant). By choosing $\alpha=0.002$, I was able to show that the loop was caused by propellant viscoelasticity during rapid pressurization (ignition transient); see figure below, where circles represent predicted history.

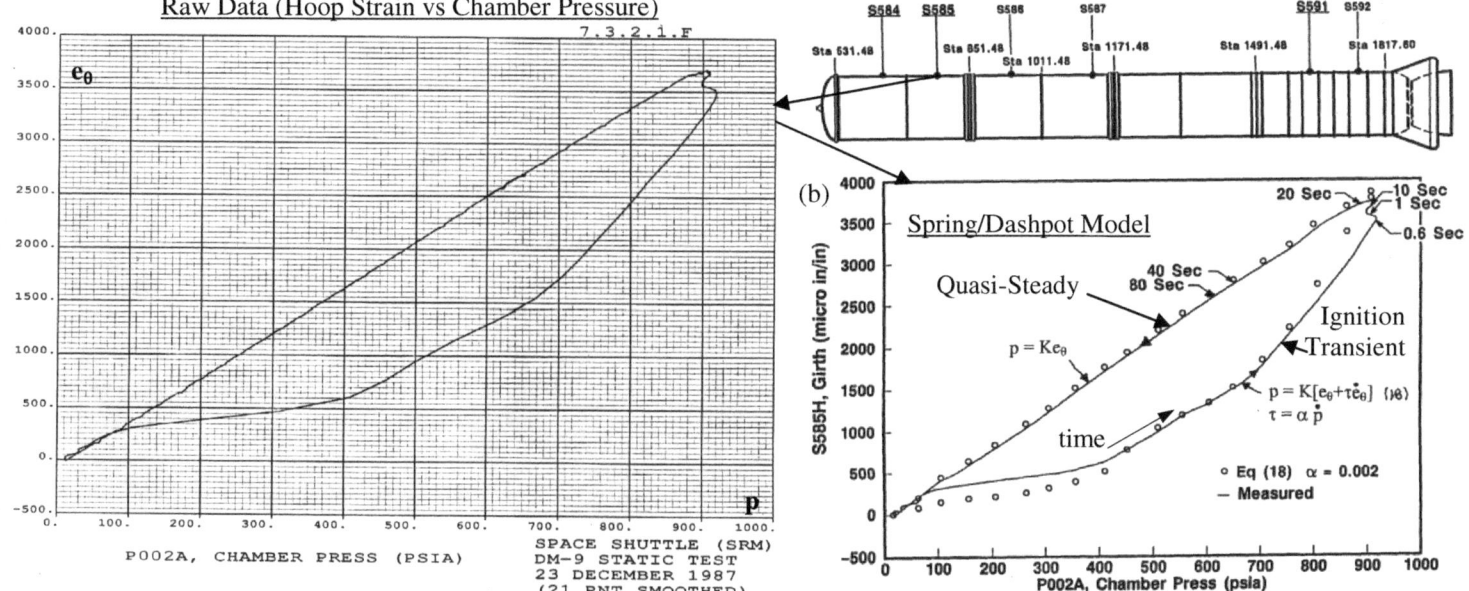

Raw Data (Hoop Strain vs Chamber Pressure)

7.3.2.1.F

P002A, CHAMBER PRESS (PSIA)
SPACE SHUTTLE (SRM)
DM-9 STATIC TEST
23 DECEMBER 1987
(21 PNT SMOOTHED)

(b) Spring/Dashpot Model

Quasi-Steady

$p = Ke_\theta$

time

20 Sec 10 Sec
1 Sec
0.6 Sec

40 Sec
80 Sec

Ignition Transient

$p = K[e_\theta + \tau \dot{e}_\theta]$ (18)
$\tau = \alpha \dot{p}$

o Eq (18) $\alpha = 0.002$
— Measured

S585H, Girth (micro in/in)
P002A, Chamber Press (psia)

A cute incident occurred at a conference where French analysts were trying to show, rather unsuccessfully, how to use hoop strain gauges to expose the motor interior pressure. During the Q&A after the presentation, I raised my hand to tell them that I had already published a paper (Ref 25) with a successful solution to their problem. They asked me for the citation information. I responded that I didn't remember it off the top of my head. Then, in the back of the room, colleague Mark Eagar stood up and said he could tell them the reference immediately because he had a copy of my paper in his hand. He further stated that "he carries it whenever he goes to a rocket meeting because it is so useful". That was an unexpected accolade.

(4G) Thermochemistry: Aids to Understanding Combustion and Ablation

Chemical Equilibrium: A Simple Closed-Form Solution Provides Much Insight

When I'm asked "what is the single most useful computer program to a rocket scientist", my answer is the NASA-Lewis Chemical Equilibrium Code CEC that was developed and expanded over a period of almost 20 years by Sanford Gordon and Bonnie McBride. It determines pressure, temperature, and species concentrations in the rocket chamber and at specified area ratios along the nozzle. It also predicts equilibrium detonation chemistry and shock jump parameters. My favorite is the next-to-last version from 1993, to which I added many additional features including (1) the ability to read ingredient properties from a master file, (2) calculation and printing of species mass fractions as well as mole fractions, (3) the ability to calculate material surface thermochemistry, (4) generation of a unified start-plane file for input to plume code SPF, (5) Victor's plume simulation with afterburning, and (6) fully-coupled plotting. I call this version **CET93**.

The Problem: However, when you see a page of CET93 output, there are numbers everywhere: mass and mole fractions for sometimes more than 30 species at possibly 10 nozzle locations. A colleague once asked me "how do you know the numbers are correct?" So I decided to construct a **closed-form solution** to this combustion problem that would provide a physically qualitative understanding and at least an approximate mathematical method to calculate the concentrations.

My Solution: I noticed that for solid propellants comprised of the usual 6 elements C, H, O, N, Cl, Al, (1) the concentrations of 6 product "**dominant species**" (CO, H_2O, HCl, H_2, N_2, Al_2O_3) always remained nearly constant throughout the chamber and nozzle and comprised at least 90% of the products, (2) the concentrations of additional 12 product "**secondary species**" (table below) comprised at least 90% of the remainder, and (3) the concentrations of all the secondary species except CO_2 disappeared as the flow moved down the nozzle. The remaining species ("tertiary") can be ignored to first approximation.

==

Subgrouping of Product Species from Most Solid Propellants

Dominant	Secondary	
CO	CO_2	AlClO
H_2O	H	OH
HCl	Cl	Cl_2
H_2	O	O_2
N_2	N	NO
Al_2O_3 (L,S)	AlCl	$AlCl_2$

==

Everywhere dominant species CO, H_2, H_2O, HCl, N_2, Al_2O_3, plus local secondary species:

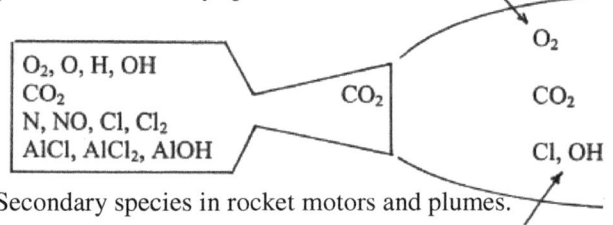

Secondary species in rocket motors and plumes.

An example for Shuttle booster propellant TP-H1148 (16% aluminum fuel, 70% ammonium perchlorate oxidizer, 14% PBAN binder) at a chamber pressure p=600 psi is shown below from the chamber to nozzle exit area ratio AE/AT=15:

INGREDIENTS (MASS AND MOLE FRACTIONS OF ELEMENT I)

		MASS FR	MOLE FR
AL	=	0.160000	0.066499
CL	=	0.211228	0.066813
H	=	0.037500	0.417211
N	=	0.087767	0.070267
O	=	0.390484	0.273690
C	=	0.113021	0.105521

$Q_k \uparrow$ Physical Exit   Demo Exit

CET93 SOLUTION

	CHAMBER	THROAT	EXIT	EXIT	EXIT	EXIT	EXIT	
P, ATM	40.827	23.550	5.3143	1.9682	0.98079	0.62779	0.36783	
T, DEG K	3389.95	3197.76	2701.68	2383.97	2313.68	2168.88	1999.05	
AE/AT		1.0000	2.0000	4.0000	7.1600	10.000	15.000	← Nozzle locations
MOLE FRACTS								
ALCL	0.00378	0.00240	0.00040	0.00006	0.00005	0.00001	0.00000	
ALCL2	0.00114	0.00075	0.00016	0.00003	0.00002	0.00001	0.00000	
ALOH	0.00034	0.00020	0.00003	0.00000	0.00000	0.00000	0.00000	
CO	0.23038	0.23137	0.23269	0.23197	0.23138	0.23037	0.22854	bold = dominant + CO_2
CO2	0.01618	0.01667	0.01865	0.02075	0.02134	0.02277	0.02492	
CL	0.01217	0.01012	0.00482	0.00213	0.00214	0.00124	0.00057	
H	0.03432	0.02785	0.01221	0.00522	0.00522	0.00297	0.00134	
HCL	0.13601	0.14164	0.15328	0.15766	0.15772	0.15900	0.15990	
H2	0.25552	0.25974	0.26978	0.27513	0.27572	0.27817	0.28104	
H2O	0.14208	0.14261	0.14345	0.14278	0.14219	0.14120	0.13937	
NO	0.00064	0.00040	0.00008	0.00002	0.00001	0.00001	0.00000	
N2	0.08177	0.08239	0.08364	0.08413	0.08414	0.08428	0.08439	
O	0.00066	0.00038	0.00005	0.00001	0.00001	0.00000	0.00000	
OH	0.00849	0.00601	0.00170	0.00050	0.00045	0.00021	0.00007	
O2	0.00015	0.00007	0.00001	0.00000	0.00000	0.00000	0.00000	
AL2O3 (A)	0.00000	0.00000	0.00000	0.00000	0.07957	0.07975	0.07986	
AL2O3 (L)	0.07385	0.07573	0.07876	0.07955	0.00000	0.00000	0.00000	

34

Warning: Chemistry and Algebra Ahead !

I have chosen to present the following approximate solution here because it is simple enough that a smart high-school senior or a college freshman could derive it, and the results provide much insight into the combustion process in a solid-propellant rocket (as well as a liquid rocket with chlorine and aluminum removed). Details are provided in Refs 1 and 28.

The secondary species other than CO_2 are the result of dissociation in the very hot combustion chamber. However, as the flow expands through the choked nozzle, the flow cools rapidly, so that all secondary species except CO_2 eventually recombine into the dominant species. Thus, only 7 species (6 dominant species plus CO_2) have significant concentrations near the nozzle exit or in the cooler non-mixing (high altitude) exhaust plume.

I realized that it was no coincidence that there are 6 dominant species and 6 atom conservation equations. Consequently, I let the mole fractions X_k of the dominant product species be constrained by the <u>atom conservation equations</u>:

	Dominant Species	Secondary Species		
$Q_C \phi =$	X_{CO}	$+ X_{CO2}$		
$Q_H \phi =$	$2X_{H2} + 2X_{H2O} + X_{HCL}$	$+ \sigma_H$	where	$\sigma_H = X_H + X_{OH}$
$Q_O \phi =$	$X_{CO} + X_{H2O} + 3X_{AL2O3}$	$+ 2X_{CO2} + \sigma_O$	where	$\sigma_O = X_O + 2X_{O2} + X_{OH} + X_{NO} + X_{ALCLO}$
$Q_N \phi =$	$2X_{N2}$	$+ \sigma_N$	where	$\sigma_N = X_N + X_{NO}$
$Q_{CL} \phi =$	X_{HCL}	$+ \sigma_H$	where	$\sigma_{CL} = X_{CL} + 2X_{CL2} + X_{ALCL2} + \sigma_{AL}$
$Q_{AL} \phi =$	$2X_{AL2O3}$	$+ \sigma_{AL}$	where	$\sigma_{AL} = X_{ALCL} + X_{ALCL2} + X_{ALCLO}$

Q_k are the mole fractions of the ingredient elements, the parameter ϕ is determined from the requirement that the sum of all mole fractions must be unity (i.e. $\Sigma X_k = 1$), and 12 equilibrium equations determine the concentrations of the 12 secondary species (i.e. σ_i) as functions only of pressure, temperature, and the concentrations of the dominant species. One example is:

$$X_{CO2} = \frac{\alpha(T) X_{CO} X_{H2O}}{X_{H2}} \qquad \text{where} \quad \log_{10}\alpha = -1.6949 + 7.9196 \times 10^{-5} T(K) + 1775/T(K)$$

The atom conservation equations can be rearranged to determine the mole fractions of the dominant species:

$$X_{H2O} = \frac{B - [B^2 - 4(1-\alpha)R_O R_H]^{1/2}}{2(1-\alpha)}$$

$$X_{H2} = R_H - X_{H2O}$$

$$X_{CO} = \frac{Q_C \phi}{1 + \alpha X_{H2O}/X_{H2}}$$

$$X_{HCl} = Q_{Cl}\phi - \sigma_{CL}$$

$$X_{N2} = \tfrac{1}{2}(Q_N\phi - \sigma_N)$$

$$X_{Al2O3} = \tfrac{1}{2}(Q_{Al}\phi - \sigma_{AL})$$

where

$$\phi = \frac{2-\sigma}{1 + Q_C - Q_O}$$

$$R_H = \tfrac{1}{2}\left[(Q_H - Q_{Cl})\phi + \sigma_{CL} - \sigma_H\right]$$

$$R_O = \left(Q_O - Q_C - \tfrac{3}{2}Q_{Al}\right)\phi + \left(\tfrac{3}{2}\sigma_{AL} - \sigma_O\right)$$

$$B = (1-\alpha)R_O + R_H + \alpha Q_C\phi$$

$$\sigma = \sigma_N + \sigma_H + X_{CL} + 2(X_O + X_{O2}) - X_{ALCL2}$$

The solution of the above equations is iterative: the mole fractions for the dominant species are estimated by first setting $\sigma_i = 0$, then the values of the secondary species and therefore σ_i are determined from their equilibrium equations, and then the dominant species are updated. This process is iterated, and convergence is rapid. For the sample case on the previous page, the first pass at the simulated nozzle exit (area ratio 15 where temperature T=1999°K → α=0.22456) yields for σ_i=0:

$$\phi = 2.40433 \quad \rightarrow \quad R_H = 0.42124 \qquad R_O = 0.16451 \qquad B = 0.60578$$

The mole fractions X_i calculated from the above equations match the numerical results on the previous page almost identically, even in this first pass. Note that atom conservation is independent of the phase of Al_2O_3 (liquid L or alpha solid A).

It is important to note that at large area ratios (1) the solution is <u>independent of pressure</u>, and (2) the solution is only <u>weakly dependent on temperature</u> through α. This explains why <u>this solution technique was successful even when CET93 failed</u>, for example at very low pressures.

> This procedure provided an improved understanding of the effect of ingredient mole fractions on combustion products, and allowed closed-form solutions to replace more-complicated numerical solutions in some flow codes. I also had cause to apply this dominant/secondary method to numerous other chemical reactions beyond rocket propellants, when numerical solutions of their non-approximated systems failed.

I also studied the 1994 Propellant Equilibrium Program (**PEP94**) from NAWC that was preferred by the gas-bag inflator community, and the Aerotherm Chemical Equilibrium Program (**ACE**) which is specialized for surface thermochemistry (ablation) calculations. I discovered several important errors in PEP94; despite notifying NAWC soon after, the errors were not fixed for several years. Likewise, I found that ACE sometimes chooses the wrong condensed species to calculate the saturation temperature. I cleaned up the confusing output, and added the important option to ACE to allow nozzle location to be defined by area ratio (surprisingly absent until then). The use of CET93 to explain ACE is described on the next page.

Surface Thermochemistry: A Conceptual Model of Material Ablation and Gas Saturation

The Problem: Heating of a surface like a nozzle wall causes the material surface to ablate, i.e. to vaporize. The resulting vapor reacts with the hot gas flowing along the wall and increases its temperature. This temperature is a boundary condition for the conduction of heat into the wall material. Consequently, we need to know its value. The Aerotherm Chemical Equilibrium code ACE has been used throughout the rocket industry to predict this near-wall temperature given the constituents and rate of ejection of the ablative vapor. Many analysts run ACE without understanding the complicated physical process, so I developed a two-pronged conceptual model: (1) I created a "salt-water analogy", and (2) showed how the easily-understood NASA-Lewis Equilibrium code CET93 can duplicate the basic ACE predictions (Refs 1 and 29), although less efficiently.

My Solution: For simplicity, assume the wall to be comprised of graphite and resin. The resin will vaporize due to heating and decomposition, but the graphite vaporizes due to chemical reaction with the oxidizer-rich gas flow adjacent to the wall. This reaction removes as much graphite (in the form of CO and CO_2) as the gas oxidation capability allows. Thus it will be "saturated" with graphite vapor, but no excess condensed graphite will appear in the gas flow. I call the resulting gas temperature at chemical equilibrium the "saturation temperature" T_{sat}, and is the minimum temperature at which no condensed graphite will appear in the flow. At higher temperatures, the only condensed graphite that can exist in the flow will be the result of structural failure and/ or removal of the graphite char by the shear of the gas flow, but not by chemical reaction.

This process can be understood using a "**salt-water analogy**": Suppose we pour a lot of salt into a beaker of cold water. Some of the salt will go into solution, but the excess will precipitate to the bottom of the beaker. If we increase the temperature of the water before pouring in the salt, more salt will go into solution and less will precipitate. If we heat the water just enough that no salt precipitates out, this is its "**saturation temperature**". Think of the graphite as the salt.

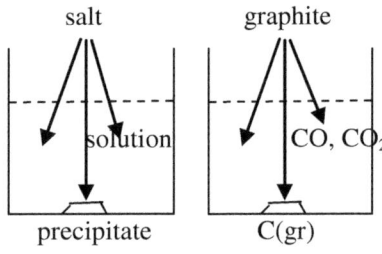

The head of the Heat Transfer group at Thiokol, Fred Perkins, came to me some weeks after I had created this analogy, and told me how very useful it was in a presentation to Air Force customers: they began to understand ablation for the first time.

The process can be duplicated using CET93: Decrease the temperature from 3500°K until any condensed phase of char or pyrolysis products begin to appear in the flow. This is then the saturation temperature. I have automated this iterative procedure by writing a utility code CETSAT.

Input:

```
* CET93 INPUT FOR ASRM PROPELLANT+ABLATIVES vs ACE: CHAR/PGAS EXPLICIT FOR BOOK
ABLATION: B'c   B'g   ◄─── Injection rate of char and pyrolysis gas
          0.1   0.1
INGREDIENTS
7
C       293    15.5        F
CL      334    32.4        F
H2      663     5.8        F
N2      864    13.0        F
O2      924    33.3        F
CHAR   1388    50.0        O
PGAS   1389    50.0        O

NAMELISTS
&inpt2  rkt=f, tp=t, p=42.529, atm=t, molmas=1,
        t=2000., 2500., 2806., 2808., 3000., 3500.,         &END
```

The saturation temperature predicted by CET93 here lies between 2806° and 2808°, which is within 12°K of the ACE solution of 2796°K.

Output: Found temperatures that bracket the saturation point

THERMODYNAMIC PROPERTIES

P, ATM	42.529	42.529	42.529	42.529	42.529	42.529
T, DEG K	2000.00	2500.00	2806.00	2808.00	3000.00	3500.00
RHO, G/CC	4.7410-3	3.7759-3	3.3499-3	3.3473-3	3.1177-3	2.6023-3
H, CAL/G	79.759	367.37	593.15	594.76	720.83	1140.80
U, CAL/G	-137.48	94.598	285.69	287.07	390.48	745.02
G, CAL/G	-5323.10	-6705.82	-7584.28	-7590.11	-8153.90	-9664.00
S, CAL/(G)(K)	2.7014	2.8293	2.9143	2.9148	2.9582	3.0871
M, MOL WT	18.295	18.213	18.136	18.136	18.046	17.573
(DLV/DLP)T	-1.00447	-1.00226	-1.00358	-1.00360	-1.00556	-1.01776
(DLV/DLT)P	1.0275	1.0234	1.0566	1.0585	1.0942	1.2775
CP, CAL/(G)(K)	0.5279	0.6530	0.8407	0.6223	0.6955	1.0160
GAMMA (S)	1.2703	1.2088	1.1654	1.2402	1.2255	1.1960
SON VEL,M/SEC	1074.5	1174.5	1224.4	1263.6	1301.5	1407.3

MOLE FRACTS

CO	0.32680	0.33181	0.34119	0.34126	0.33957	0.33064
C2H2	0.00037	0.00516	0.01602	0.01611	0.01536	0.01233
CL	0.00004	0.00055	0.00185	0.00186	0.00341	0.01167
H	0.00016	0.00243	0.00814	0.00820	0.01530	0.05545
HCN	0.00242	0.01134	0.02166	0.02173	0.02230	0.02200
HCL	0.13258	0.13376	0.13620	0.13621	0.13397	0.12205
H2	0.41905	0.41925	0.41372	0.41367	0.40958	0.38469
N2	0.06610	0.06251	0.05921	0.05919	0.05852	0.05651
C(GR)	0.04813	0.03157	0.00024	0.00000	0.00000	0.00000

Saturation Point

Schematic of Material Ablation

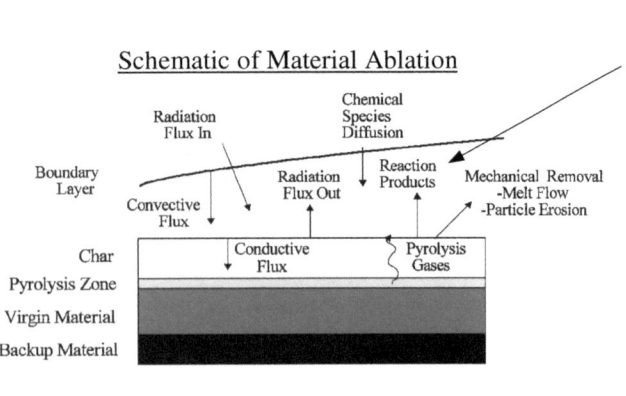

Hybrid Rockets

As opposed to the solid propellant rocket motor whose fuel and oxidizer are pre-mixed and solidified, the hybrid rocket motor utilizes separate liquid oxidizer and solid fuel. The advantage of the hybrid motor is that it can use propellants that generate less-noxious combustion products and is safer because the combustion can be terminated at will by shutting off the oxidizer.

Hybrid rockets were studied in some depth in the 1960s, but thrust magnitudes were too small to be practical. This was due primarily to the low regression-rate at which the solid phase recedes. Consequently, hybrid motors often require multi-port fuel grains (see figure below), which have poor volumetric efficiency and, often, structural deficiencies.

A funny story was told by Prof. David Altman of Stanford: A scale model vehicle with hybrid motor was built using salami as the fuel. After launch, the vehicle flew over a nearby hill. The launch team recovered the vehicle. In order to demonstrate the environmentally safe nature of the hybrid propellant, the team ate the unburned remnant ("sliver") of salami.

Partly as a result of the motor failures discussed herein (Challenger on pages 31-32, Titan SRMU and SICBM on page 44), renewed interest in hybrid rockets began in the late 1980s. It was hoped that new materials and technologies would allow larger hybrid boosters to be constructed. As part of this effort, I was asked to model the hybrid combustion process. I worked this project for about a year with Thiokol colleagues Golafshani and Loh, and published two papers (Refs 39, 40).

Schematic of Hybrid Rocket Motor

1 Injector Preburner — **2 Solid Fuel Grain** — A — **3 Mixer** — **4 Nozzle**

oxidizer

Liquid Oxidizer

A

Section A-A

- Oxidizer vaporization
- Oxidizer distribution and port initial conditions

- 3-D, two-phase, turbulent, reacting, radiating flow
- Fuel decomposition chemistry and fuel surface boundary conditions

- Fuel/oxidizer afterburning
- Multiport flow mixing

- 3-D 2P, turbulent, reacting, radiating flow

Shortly after this effort, I left Thiokol and was never again involved with hybrid rocket motors. Indeed, the national effort to develop hybrid motors as boosters petered away due to the inability to generate sufficient thrust and impulse to allow hybrid motors to compete with pure liquids or solids. Nonetheless, high regression rate liquefying fuels developed in the late 1990s offer a potential solution to this problem.

Summary of My "Table-Top" Experiments

Assumptions often must be made when modeling complex phenomena in and around rockets. In order to validate the accuracy of those assumptions, experiments must be conducted that isolate the phenomena. Full-scale tests are often too expensive or incur coupling with other phenomena. Consequently, it is desirable to formulate an experiment that can be conducted simply and inexpensively on a work bench or table top. Some 'table-top' experiments that I designed in several days and conducted within a few hours with help from lab personnel to validate assumptions in my modeling are discussed in these memoirs:

- Resin Casting to Simulate Propellant Batch Interface (page 18)
- Greased O-Ring for Static Coefficient of Friction in Groove (page 95)
- Collision-Coalescence of Mercury Droplets (page 40)
- Surface Impact of Mercury Droplets (page 41)
- Slumping Mercury Droplets (page 41)

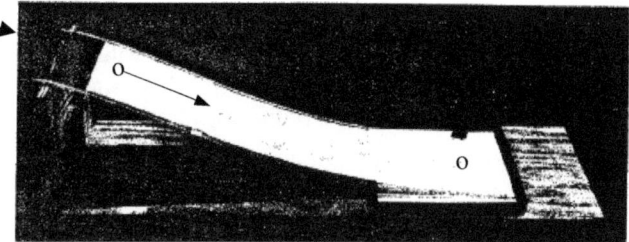

"Ski Slope" (2 ft long) for collision/coalescence tests: droplet placed at top rolls to impact stationary target droplet at bottom … see page 40 for results.

(4H) Dynamics of Liquid Metal Droplets

Log-Normal Distributions

Warning: Math Ahead !

The Problem: The distribution of many properties in the rocket business is found to be **log-normal**, i.e. plots as a straight line on log-probability paper. However, for many of the log-normal calculations I had to make (integrals, etc), it was advantageous to be able to describe a log-normal distribution mathematically. I first asked the engineers in the industry who were well known for their use of log-normal functions, but they didn't know how to formulate them mathematically. I then went to our statistician and asked him if he knew how. His answer was "**I'm a statistician, not a mathematician.**"

My Solution: Well, I guessed that I'd have to figure it out myself. So I went to my copy of <u>Handbook of Mathematical Functions</u> (Ref 78), always a font of information. I looked in the index under "probability function" and scoured through pages of relationships. And there it was: the probability function written in terms of the well known "error function" erf. Consequently, the mass fraction of a log-normal distribution of particles accumulated between diameters of 0 and D is

$$\boxed{f(D) = \frac{1}{2}\left[1 + \text{erf}(Z)\right]} \qquad \text{so that} \qquad \frac{df}{dD} = \frac{\exp(-Z^2)}{(2\pi)^{1/2} D \ln\sigma_g} \qquad \lambda = \ln\sigma_g \qquad (10)$$

where

$$Z = \left[\log_{10}\left(\frac{D}{D_m}\right)\right]\frac{1}{\sigma\sqrt{2}} \qquad\qquad D_m = \text{mass-median diameter (where } f = \frac{1}{2})$$

$$\sigma = \text{standard deviation} = \log_{10}\sigma_g \qquad\qquad \text{erf} = \text{classical error function}$$

The corresponding number distribution accumulated for a log-normal mass distribution is

$$g(D) = \frac{1}{2}[1 + \text{erf}(Y)] \qquad \text{where} \qquad Y = Z + 3\frac{\lambda}{\sqrt{2}}$$

> This mathematical description of log-normality was to be a breakthrough in comparing distributions documented by many different researchers or described in different disciplines.

These relationships allow many other properties to be calculated in closed form (Refs 15, 28). For example, the sieve specifications for 200µm ground AP (Ammonium Perchlorate oxidizer) in Shuttle booster propellant are equivalent to two log-normal distributions:

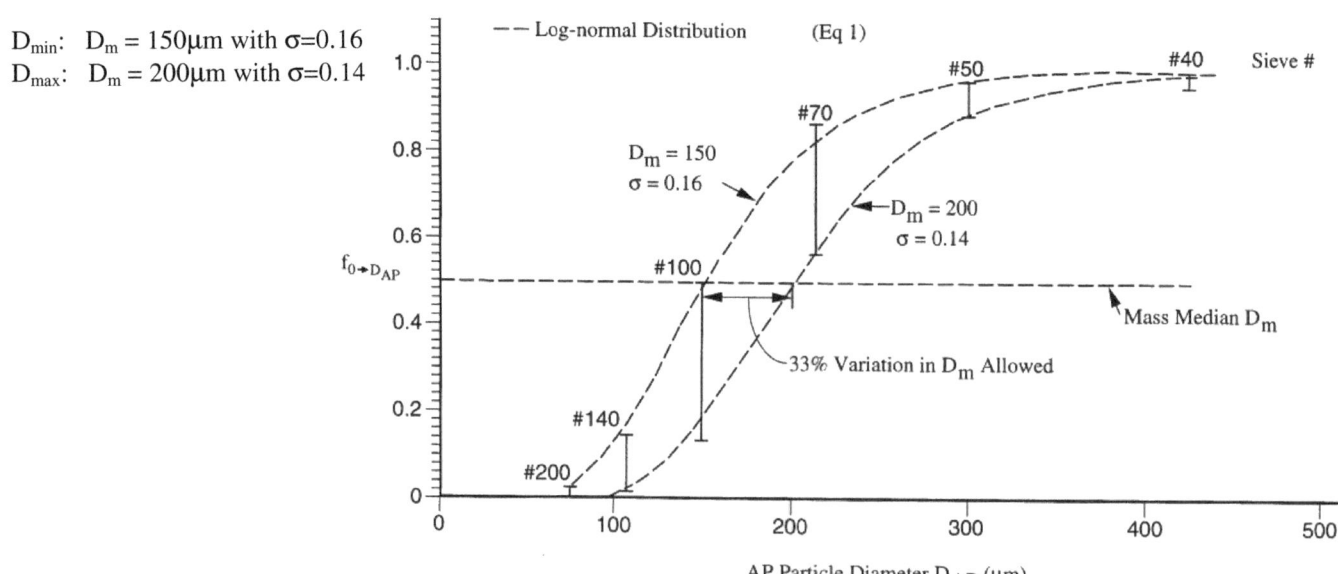

Sieve Specifications for Ground AP for Shuttle Boosters
(Note excellent curve-fit by log-normal functions eq(10))

D_{min}: $D_m = 150\mu m$ with $\sigma = 0.16$
D_{max}: $D_m = 200\mu m$ with $\sigma = 0.14$

AP Particle Diameter D_{AP} (µm)

I encountered another example of the usefulness of this error function formulation. A report from Allegany Ballistic Labs (ABL) spent many pages documenting and validating the numerical (quadrature) solution of an integral of $D^k df/dD$ as part of a combustion study. I was able to show that using eq(10), the integral had an exact solution (see Appendix C of Ref 1). This would have saved ABL much effort and potential error.

Application of log-normality to the size distribution of aluminum oxide droplets formed during combustion of aluminized rocket propellants is described in the following pages.

Bimodal Size Distribution of Al_2O_3 Droplets

The Problem: The first static firing of a Shuttle booster (ETM-1A) after the Challenger accident generated 50% more aluminum oxide slag than the average from the previous set of motors. ETM-1A was the oldest booster yet fired, so the influence of propellant age on slag generation came into question. Since slag depends strongly on the size distribution of the Al_2O_3 droplets formed during combustion, it was decided that we needed to learn more about the actual size distribution by burning propellant samples in the same Quench Particle Combustion Bomb (QPCB) used previously for other booster propellants. The sample was ignited at a distance "s" from a liquid layer, and the ejected aluminum oxide droplets would penetrate the liquid and freeze (be "quenched").

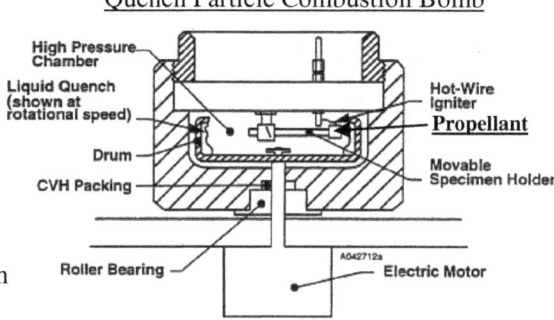

Quench Particle Combustion Bomb

The resulting mass distribution of the quenched particles as measured by a MicroTrac particle analyzer was given to me.

My Solution: I observed that contrary to long-held assumptions, **the mass distribution is bimodal**, with 65-80% of the mass in the form of "**smoke**" whose mass median diameter D_{ms} was always around 1.5μm, and the remaining 20-35% of the mass having "**coarse**" mass-median diameters $D_{mc} \approx 180$μm at 50 psi down to 90μm at 1000 psi (Refs 1, 30). Each mode is created by different physical mechanisms. When the aluminized propellant burns, the aluminum powder melts and is blown off the surface as molten "agglomerates" by the gaseous oxidizing species. The aluminum agglomerates oxidize at a finite rate as they flow toward the nozzle. Smoke is the result of the reaction of the oxidizing species with gasified aluminum in a diffusion flame around each aluminum droplet, while the coarse droplets are the oxidized remainder of the original aluminum droplet.

> I was able to curve-fit each data mode with a separate log-normal distribution (see figures below). This mathematical model was another breakthrough in understanding Al_2O_3 droplet size distributions generated from all aluminized propellants.

Measured Bimodal Size Distribution of Al_2O_3 from TP-H1148   Accumulated Mass of Al_2O_3 (Bi-modal log-normal vs measured)

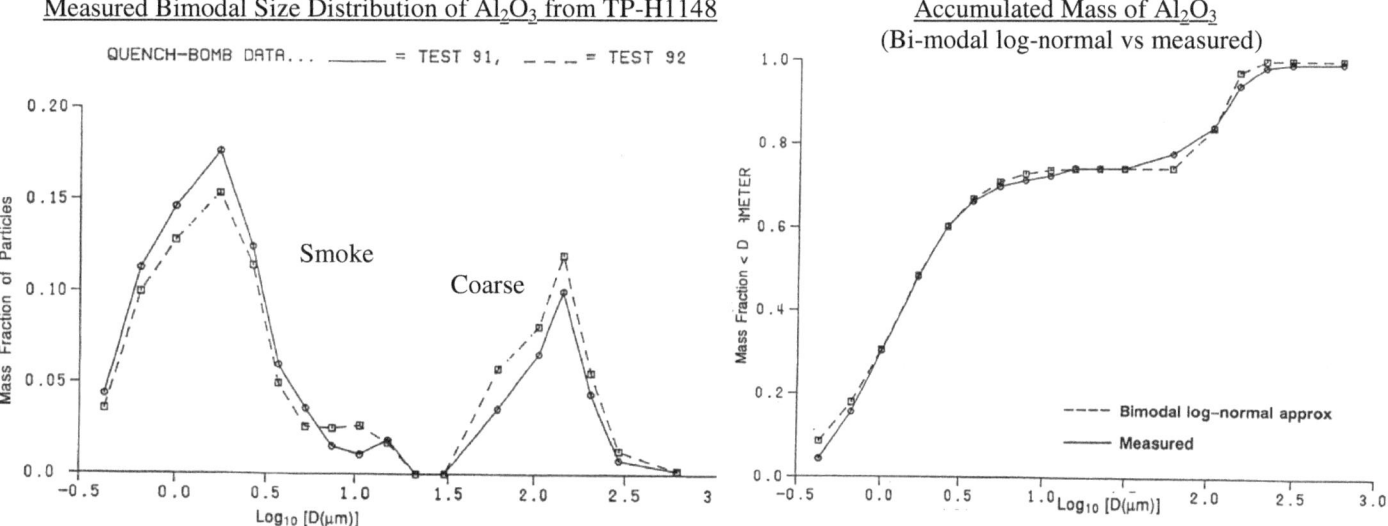

Sometimes, the bimodal character of the droplet distribution is not obvious when the two modes overlap (see below); nonetheless, a bimodal log-normal curve-fit with smoke $D_{ms}=1.3$μm and coarse $D_{mc}=13.3$μm still matched the size distribution well:

Measured Bimodal Size Distribution of Al_2O_3 from ANB-3088   Accumulated Mass of Al_2O_3

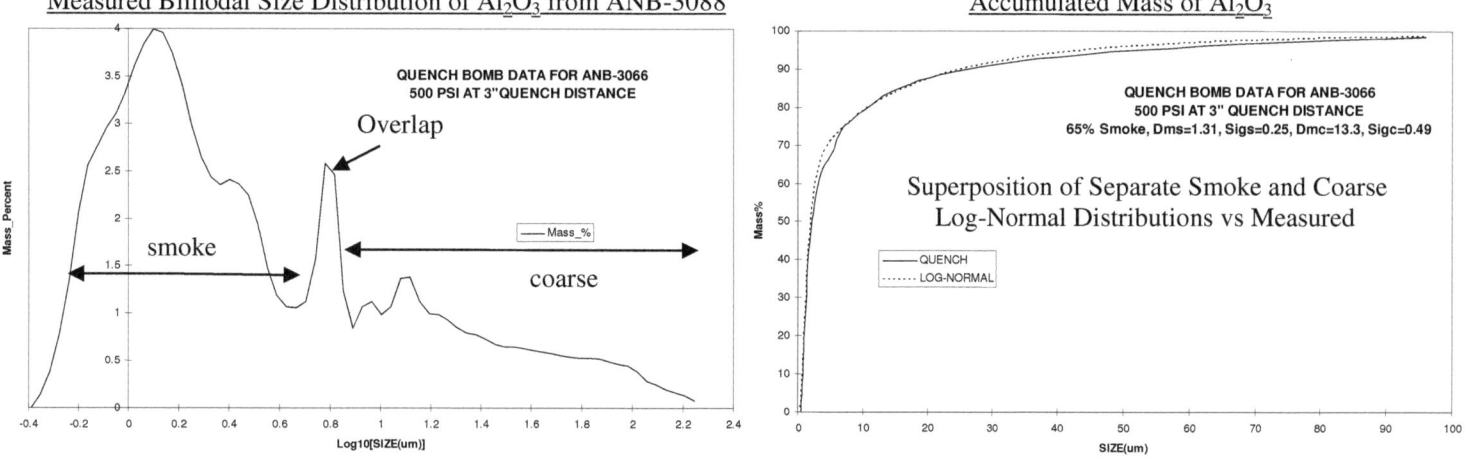

Droplet Dynamics – A Model of Collision/Coalescence of Al_2O_3 Droplets Using Mercury

The Problem: As shown above, combustion of aluminized solid propellants produces aluminum oxide droplets in a wide range of sizes in the combustion chamber. When they enter the nozzle, droplets of different sizes accelerate at different rates and may collide. Early modeling assumed that all collisions among these droplets resulted in coalescence. However, I found a model for coalescence efficiency previously developed and verified experimentally for free-falling water droplets that showed that permanent coalescence will not occur if the rotational energy of the spinning (temporarily coalesced) agglomerate (due to collision offset x) exceeds the surface energy holding it together. Would this also apply to Al_2O_3 droplets?

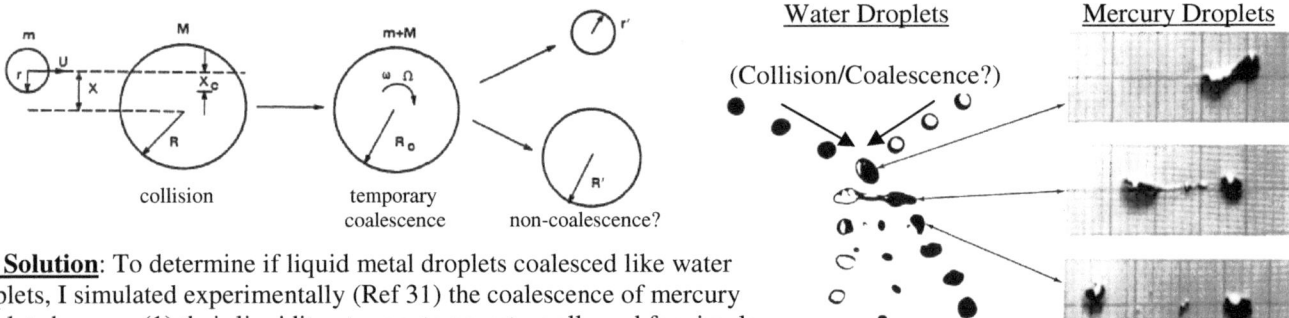

My Solution: To determine if liquid metal droplets coalesced like water droplets, I simulated experimentally (Ref 31) the coalescence of mercury droplets because (1) their liquidity at room temperature allowed for simple and inexpensive testing, (2) they have high surface tension and density that is comparable to that of Al_2O_3, (3) their large diameters are easily visible, and (4) their high density generates high-impact energy at low-impact velocity. Single mercury droplets ("bullets") of various diameters were rolled down a "table-top ski slope" (p37) into a stationary mercury droplet ("target"), and the collision/coalescence process was recorded on videotape; bullet size, speed, and offset were deduced from a superposed grid. It was concluded that <u>the water model accurately predicts the collision-induced coalescence of mercury, and therefore Al_2O_3, droplets</u> because (1) the qualitative similarity between the collision/coalescence behavior of water droplets and liquid metal (mercury) droplets was excellent (figure above), (2) the model predicted well quantitatively whether a collision resulted in coalesced mercury droplets, and (3) the model predicted well the degree of Al_2O_3 coalescence apparently occurring in the nozzle of the Space Shuttle booster (see next section).

Droplet Dynamics Through the Nozzle – OD3P

The Problem: The One Dimensional 3-Phase CFD code (OD3P) was written by Jim Kliegel's think tank KVB to model the behavior of metal droplets in 1D multiphase flow through a rocket nozzle. The code models 5 size-change mechanisms: (1) aerodynamic shattering, (2) collision/coalescence with other droplets, (3) cooling-induced shrinkage, (4) condensation, and (5) solidification. It was a nice piece of work, but it had some coding errors, and the predicted mass-median diameter of the droplets at the nozzle exit of the test case overpredicted the value from Hermsen's correlation of experimental data.

My Solution: I found and fixed the coding errors, and determined that the cause of the inaccurate prediction was primarily due to the assumption that all droplet collisions result in coalescence, which they don't. When I coded the water/mercury model for collision/coalescence described above into OD3P, the results (Ref 32) showed that

(1) <u>collisions between large droplets rarely resulted in coalescence</u> (see pictures above), while
(2) <u>collisions of smoke droplets with large (coarse) droplets always resulted in coalescence</u>, and
(3) the agreement of OD3P-predicted diameter distribution with Hermsen's correlation and with the measured distribution at the nozzle exit was greatly improved.

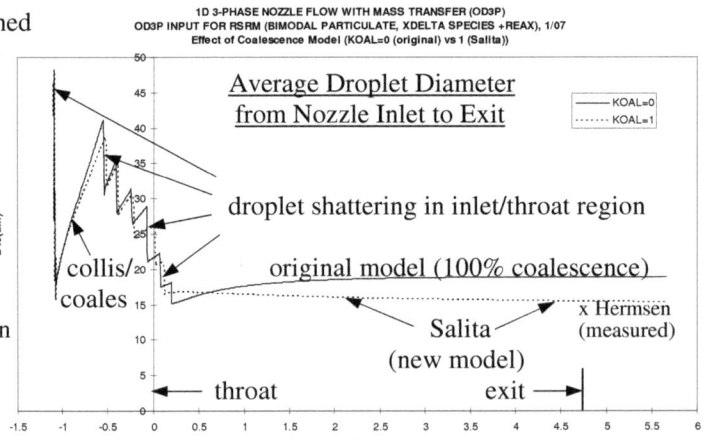

These results had a major impact on the understanding of the transition of the size distribution of Al_2O_3 droplets from the combustion chamber through the nozzle and into the exhaust plume (Ref 33). In particular,

(1) the coarse mode of roughly 20-35% 100μm droplets in the RSRM chamber shattered in the nozzle inlet to roughly 1000 times as many 10μm droplets in the nozzle inlet,
(2) the smoke mode of roughly 65-80% 1.5μm droplets collided and coalesced with these shatter products, resulting in
(3) a <u>mass</u> distribution at the nozzle exit that is roughly 98% shatter products and only 2% smoke. However,
(4) the <u>number</u> of coarse and smoke droplets at the nozzle exit and into the plume are now of the same order.

As a result, the plume community was forced to revise their models of plume radiation, which depend strongly on particle size. The above predictions were later supported by Sambamurthi when he collected and measured particulate on sticky darts fired through the exhaust plume from an RSRM during static firing (Ref 34).

Droplet Dynamics – Surface Impact and Slump of Liquid Metal Droplets

The Problem: It is desirable to know what happens when an Al_2O_3 droplet hits a nozzle wall. Does it bounce, stick, or shatter? What happens if it hits an existing film of aluminum oxide? The behavior affects slag retention, film build-up along a nozzle throat, or impingement erosion along the nozzle wall.

Our Study: The success of the mercury collision/coalescence project suggested the possibility that mercury droplets could also help in understanding the dynamics of Al_2O_3 droplet/wall interactions. In an attempt to learn more about the impact of Al_2O_3 droplets with nozzle surfaces, we dropped mercury droplets from various heights and over a range of angles onto a variety of nozzle materials. Droplets dropped from small heights remained intact upon impact, simply shivering like jello and remaining in a slumped configuration. As the drop height was increased, more and more small "satellite" droplets were shed from the core droplet upon impact. Only at large drop heights (high impact velocity) did the core droplet shatter into several large droplets. I had insufficient funding to pursue a model of the satellite formation or the shattering process (caused by shock waves bouncing around inside the core droplet). However, I did generate an interesting model of the slumping droplet.

I wrote a computer program **DROPLET** to predict the shape of a slumping droplet where the body force due to gravity was balanced by its surface tension (Ref 35). In order to validate the predictions, we ran a short "table-top" experimental program:

We placed three mercury droplets (small, medium, large) of initial diameter D_o on a flat surface and photographed them from the side. Their shapes were predicted by DROPLET as a function of the shape parameter S:

$$S = \frac{\rho g D_o^2}{2\sigma}$$ ρ = mass density

A near-perfect match of predicted shape to each photograph was obtained by input of an appropriate value of S as long as the surface tension $\sigma(T)$ for mercury was evaluated at a temperature much higher than room temperature due to the **radiative heating from the photo lamps**.

The reflection of the photographer's lights off the mercury surface made the droplet look like a flying saucer. It's a good show stopper.

Similarity of Predicted and Measured Droplet Shapes
Predicted by DROPLET:

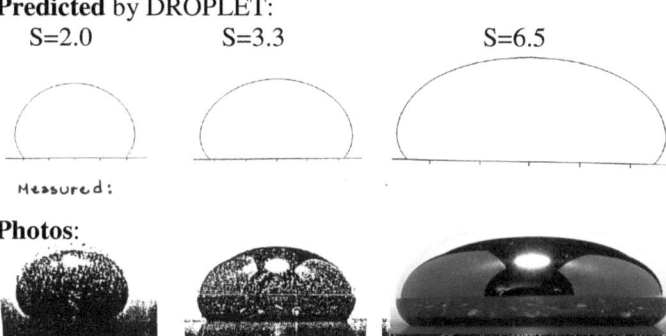

S=2.0   S=3.3   S=6.5

Measured:

Photos:

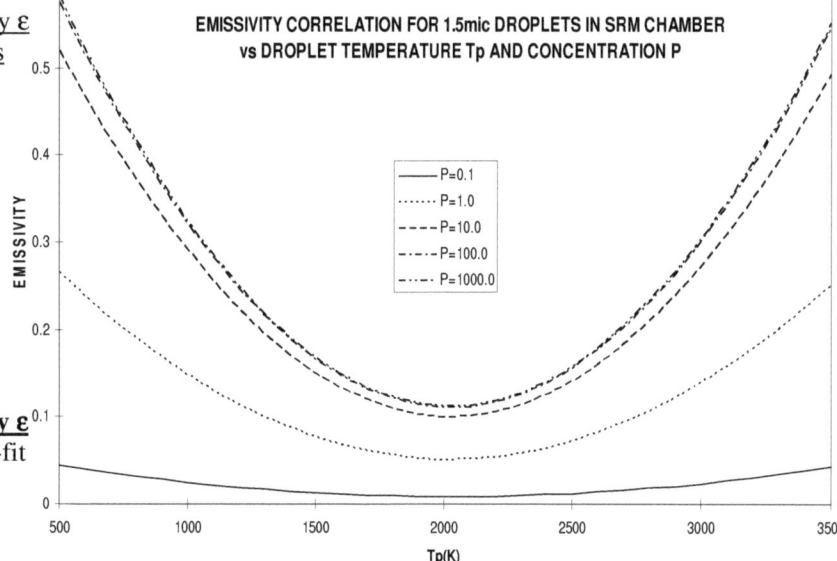

Landed UFO?

Droplet Radiation in the Combustion Chamber

The Problem: Radiative heat transfer can be as important as convective heating in the combustion chamber of rocket motors with aluminized propellants. It affects the onset of propellant ignition, and the heating and ablation of chamber insulation. However, radiative heating was usually ignored by analysts because it was strongly dependent on ill-defined droplet size distributions, temperature T_p, and indices of refraction that can be distorted by impurities (e.g. soot or unburned aluminum). The resulting equation for radiative flux depended on 12 parameters.

My Solution: Now that we knew the size distribution of Al_2O_3 droplets in the rocket chamber, I realized that we could estimate the radiative flux $Q=\varepsilon T_p^4$ from these droplets for the first time. Noting that Mie theory shows that radiation from each droplet mode is proportional to its mass fraction f divided by its mass-median diameter D_m, then the radiative flux would be dominated by the smoke since its $f/D_m=0.65/1.5=0.44$ is much larger than that for coarse droplets $0.35/100=0.0035$. And since soot or unburned aluminum rarely appears in smoke droplets, the equation for <u>cloud emissivity ε could be simplified to depend on only 2 parameters instead of the original 12</u>: droplet temperature T_p, and a single concentration parameter P :

$$P = 0.75 \, \rho_{ratio}(T_p) \, r_{ratio} \, f_s \, \Phi$$

where ρ_{ratio} is the ratio of flow mixture density to droplet bulk density, r_{ratio} is the ratio of chamber radius to droplet radius, f_s is the mass fraction of Al_2O_3 that is smoke, and Φ is the mass fraction of Al_2O_3 in the flow. The optical coefficients for absorption and scattering were determined from Mie theory as functions of droplet temperature T_p and concentration P. The **hemispherical emissivity ε** of the droplet cloud was then calculated and curve-fit quadratically (see figure at right).

Results were documented in Ref 36.

EMISSIVITY CORRELATION FOR 1.5mic DROPLETS IN SRM CHAMBER
vs DROPLET TEMPERATURE Tp AND CONCENTRATION P

P=0.1
P=1.0
P=10.0
P=100.0
P=1000.0

EMISSIVITY

Tp(K)

(4I) Two-Phase Flow and Slag Generation

The Problem: In order to shorten a motor without removing propellant, the nozzle inlet is often "submerged" into the combustion chamber. However, this submergence creates a cavity behind the nozzle where Al_2O_3 droplets can get entrained in a "recirculation zone" to form a pool of trapped Al_2O_3 called "slag" (see figure below). This causes a loss in performance of the motor, and can also result in slag "sloshing" or asymmetric ejection, which can lead to undesired nozzle torque.

In the late 1980's, two static firings of motors with submerged nozzles generated unusually high slag weights: the Shuttle booster ETM-1a and a lengthened ("37K") first-stage SICBM motor. Unfortunately, no two-phase CFD codes existed at that time to model the gas/droplet flow in motors with submerged nozzles to help understand the cause of the high slag.

SICBM Stage 1 Short and Long Motors

Submergence Cavity

Slag Accumulation
Behind Submerged Nozzle

Our Solution: My colleagues Mehdi Golafshani and Roy Loh at Thiokol developed a robust two-phase CFD code called SHARP that would predict the trajectories of Al_2O_3 droplets of specified diameters through the combustion chamber and nozzle. My first task was to tell them which droplet diameters to utilize. I assumed that smoke droplets would follow gas streamlines and easily exit through the nozzle, but that some larger and more-outboard coarse droplets would be trapped as slag. Using the RSRM and SICBM quench bomb data, I calculated those N=9 diameters that would represent equal droplet mass fractions $f_n = (2n-1)/2N$ of the log-normal distribution eq(10) of coarse droplet mass (Ref 28). After Mehdi and Roy ran SHARP with these diameters at multiple burn times, I calculated the rate of slag mass retention assuming several different capture rules. Integration of these retention rates over motor burn time yielded the predicted slag weight. A team effort.

Flow simulations were made at 6 burn times. Results are shown below for the aft end of the Shuttle booster at 70 sec burn time: computational grid, gas flowfield, and trajectories for one of the 9 droplet sizes ($D_m = 100\mu m$). The resulting entrapped slag accumulation integrated over the full 120 sec of burn over all 9 droplet sizes is shown to nearly equal the 2077 lbm average weight measured for all 14 HPM Motors (Ref 4); however, a $D_m = 140\mu m$ would be required to match the weight measured in ETM-1a. That discrepancy was never resolved.

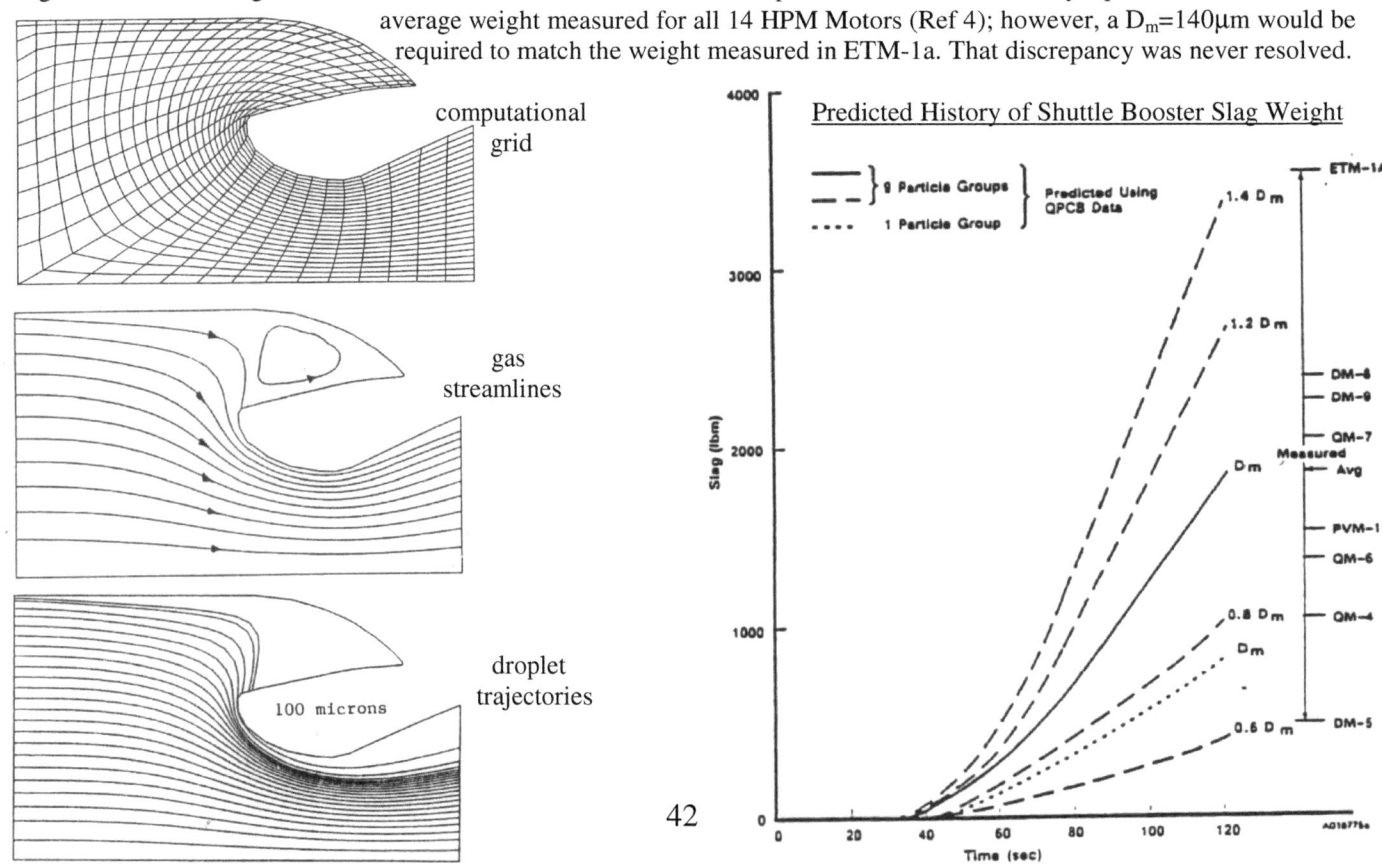

computational grid

gas streamlines

droplet trajectories

100 microns

Predicted History of Shuttle Booster Slag Weight

The SICBM Problem: The slag weight generated in static tests of the original "short" first stage of the Small ICBM (SICBM) was fairly small. However, when the Air Force requested that the motor be redesigned to generate larger impulse (time-integral of thrust), the motor case was elongated and the propellant annulus was lengthened. Surprisingly, when the longer motor was static tested, it was found that the amount of slag accumulated in the submergence cavity had <u>tripled</u> for unknown reason to one of the largest per capita weights previously experienced in a large booster.

Our SICBM Solution: In an attempt to explain this increase, Roy and Mehdi ran SHARP to simulate the gas/droplet flow-field at four burn times, using the nine diameters of Al_2O_3 droplets that I determined would simulate the measured log-normal distribution (Ref 38). The static test had been conducted in a vertical position, and the predicted slag weight calculated with an axial acceleration of 1.0 g (gravity) matched the measured slag weight pretty closely <u>only if all droplets impinging on the backside of the nozzle were also assumed to be captured as slag, in addition to those entrained in the recirculation zone</u>. The flowfield and slag calculations were repeated assuming a flight acceleration history (5.8-8.9g), with minimal change in predicted slag weight (consistent with the observation that the predicted droplet trajectories changed little with the simulated accelerations).

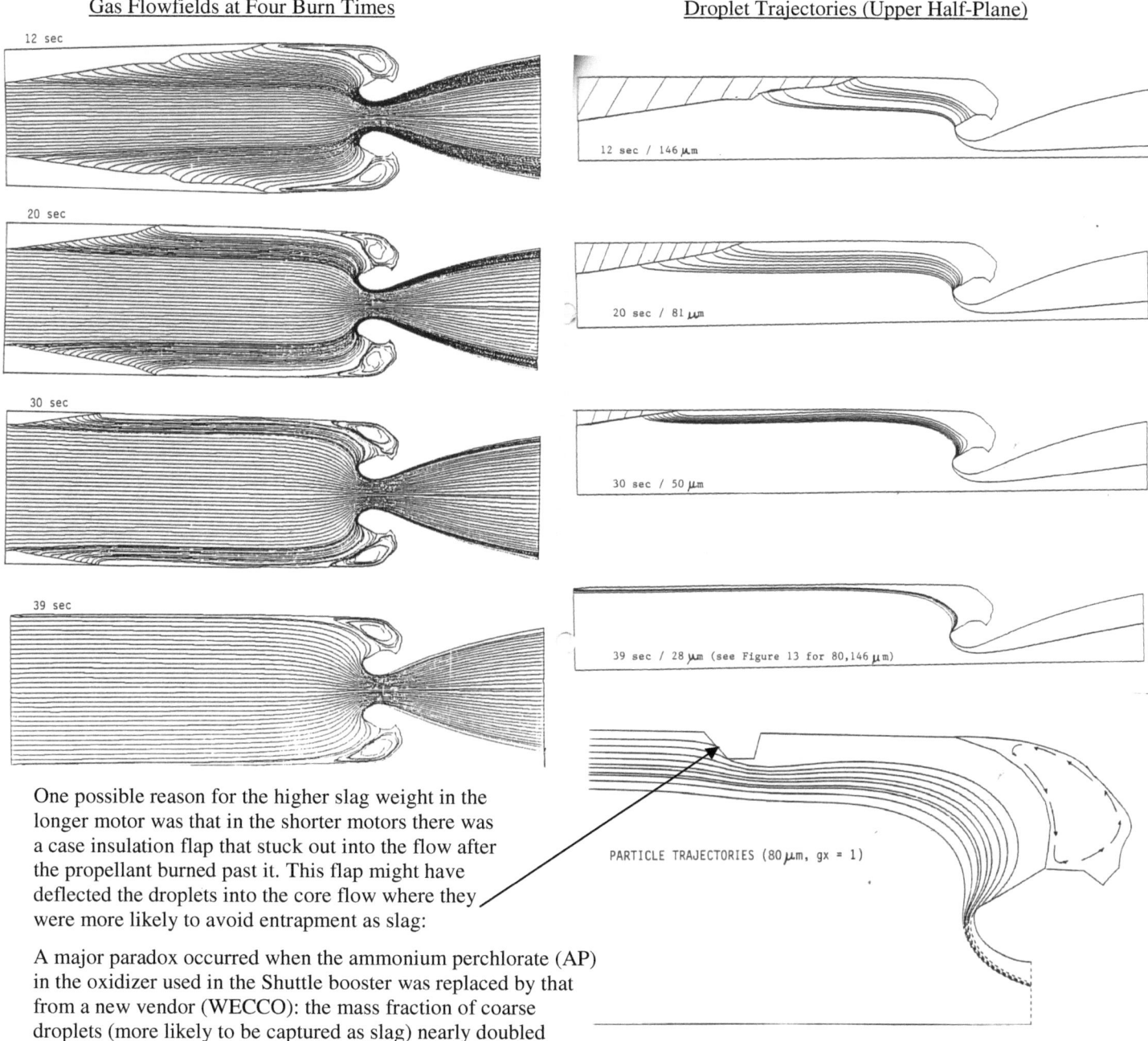

Gas Flowfields at Four Burn Times   Droplet Trajectories (Upper Half-Plane)

12 sec

20 sec

30 sec

39 sec

12 sec / 146 μm

20 sec / 81 μm

30 sec / 50 μm

39 sec / 28 μm (see Figure 13 for 80,146 μm)

PARTICLE TRAJECTORIES (80 μm, gx = 1)

One possible reason for the higher slag weight in the longer motor was that in the shorter motors there was a case insulation flap that stuck out into the flow after the propellant burned past it. This flap might have deflected the droplets into the core flow where they were more likely to avoid entrapment as slag:

A major paradox occurred when the ammonium perchlorate (AP) in the oxidizer used in the Shuttle booster was replaced by that from a new vendor (WECCO): the mass fraction of coarse droplets (more likely to be captured as slag) nearly doubled in quench bomb tests, but the slag weights measured after static tests roughly halved. This paradox is still not understood.

I was to spend much more time studying slag generation in SRMs after moving to TRW, as discussed in Chapter 5. So much so that I became known as "**Mr. Slag**".

(4J) – Motor Failures

Titan SRMU Detonation on the Pad: Effect of Propellant Overhang

The Problem: In segmented solid-propellant rocket motors (e.g., page 17), the bore of the propellant annulus is tapered, primarily for easy removal of the mandrel after casting and curing. The thick end of the downstream segment always faces forward so that the flow cross-sectional area will increase in the downstream direction in order to limit the increase in bore Mach number (by Bernoulli's Law) during motor burn.

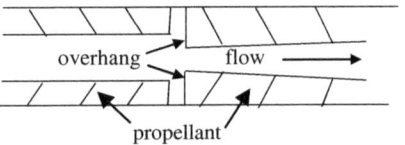

In the case of the Space Shuttle booster and Titan SRMU, this causes the forward face of the downstream segments to protrude ("overhang") into the bore. That protrusion acts like a forward-facing step, against which the gas flow in a burning motor stagnates, so that the pressure there is high. In addition, in both those motors, at least one face of each slot between segments is often non-inhibited, so that they are burning and injecting flow out into the bore flow. This causes the pressure behind the overhanging corner (point 2 in the figures below) to drop. The combination of these two effects causes a pressure torque tending to pull the propellant overhang into the bore.

My Solution: Since the Shuttle booster (RSRM) had three overhangs, I modeled semi-empirically their effect on the pressure distributions around the corners by approximating the real configuration as the sum of (1) cylindrical flow past a burning slot without overhang, plus (2) flow past the overhang without slot flow.

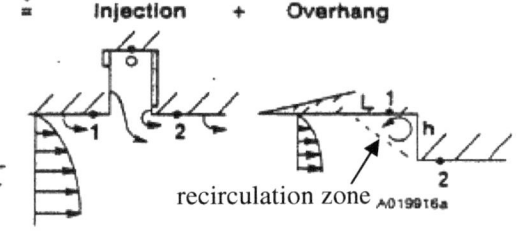

An analytical solution for the former configuration was available from Fred Culick (see page 72), and measured pressure data for the latter was available from tests with forward-facing steps. The resulting comparisons to both CELMINT and PHOENICS CFD predictions were good, and explained why the pressure differentials around the overhanging RSRM corners were acceptable (small overhang and high propellant modulus).

This solution was published in the 27th Aerospace Sciences Meeting (1989, Ref 25). Two years later, the first static test of the Titan SRMU exploded at motor ignition. The cause of this failure was subsequently identified: in the original SRMU design, the overhang was too large for its low-modulus propellant. As a result, during motor ignition the high near-stagnation pressure on the upstream-facing overhang and low pressure downstream of the sharp corner (point 2 in the above figures) caused the front of a segment in the first static test to collapse into the bore, blocking flow out the nozzle, causing huge chamber overpressure and subsequent detonation of the motor and destruction of the half-billion dollar test stand at AFRL.

> **Lesson Learned**: Had the designers of SRMU done a proper engineering job, they would have (1) found and used my simple correlation to identify that their overhang was too large, and (2) found the fix ("Ritchie Rounding") for this problem previously published by Glick/Caveny/Thurman after a similar failure of a Castor II motor at ignition.

SICBM Flight Failure: Believed Due to Slag Ejection

The first flight test of the SICBM became uncontrollable just after staging of the first stage (1989). Jim Kliegel, a well-respected engineer and department head from TRW, believed the loss of control resulted from asymmetric slag ejection through the stage 1 nozzle due to boiling of its large slag pool (as discussed on page 54) during pressure "tailoff" at staging, thereby causing the stage 1 head-end to pivot and damage the nested second stage nozzle before stage separation (see figure on page 24). The range safety officer subsequently destroyed the rest of the vehicle. More about this later.

I had gotten to know Jim Kliegel as TRW's representative to the JANNAF PSS when I was Thiokol's rep. After the SICBM flight failure, there was a meeting that we both attended to investigate the accident. He recommended several computational studies that should be conducted. I showed him and the other attendees that Thiokol had already carried out those studies, and presented the results of my slag accumulation calculations (Ref 38). My preparedness impressed Jim so much that when Thiokol laid me off in a Reduction in Force (RIF) three years later (see next page), he asked me to come work for him at TRW/San Bernardino (**Example #5 of how you never know when a document you wrote will be critical to your future**). It took until June 1993 to get the employment details worked out.

As a result of this proposed failure, the SICBM first stage motor was redesigned with a de-submerged nozzle with the expectation that the slag would be eliminated. However, I warned that the chamber aft end would now look like a forward-facing step, which would create a recirculation zone (see figure above) that would still entrap some slag aerodynamically. Indeed, during the first static test of the new motor, enough slag was generated there that it flowed into the "splitline" between the ball and socket of the nozzle. Partway through the motor burn, this slag cooled and solidified, which prevented the nozzle from vectoring. Not long after, the SICBM project was canceled due to constraints of the START treaty with the Soviets.

> Ironically, it was to be another flight failure (Orbital's Pegasus XL) attributed to slag ejection that was later to get my job back at TRW in 1994 (see page 53).

Technion Lectures (1985-1995)

I met an Israeli professor from the Technion at an AIAA conference in 1984. When I told him that my wife and I would be coming to Israel the next year to visit my parents who had retired to Jerusalem, he offered to arrange funding to pay for my air fare if I would present an all-day lecture at the Israel Institute of Technology (Technion) in Haifa. I said that I'd be honored. That arrangement in 1985 was repeated in 1989, 1992, and 1995. Subsequently, I was to teach an entire semester in 2009 (see Chapter 7).

הטכניון – מכון טכנולוגי לישראל

הפקולטה להנדסה אוירונוטית

Technion — Israel Institute of Technology

Department of Aeronautical Engineering

ס מ י נ ר י ו ן

Lecture Series

Selected Topics in Analytical Modeling of Flows in Solid Rocket Motors

by

Mark Salita
Morton Thiokol Inc.
Brigham City, Utah
U.S.A.

The Ogden Wind Quintet Plays for Thiokol

Thiokol Spokesman Gil Moore and his wife Phyllis were good friends of ours. Gil asked me to have my Wind Quintet play for his retirement party at the Thiokol Management Training Center. We happily did. After retirement from Thiokol, Gil taught at the Air Force Academy for several years, with the honorary rank of General. On one of our visits, he took us on a tour of the Academy. It was impressive the way the young Cadets would snap to attention and salute as we passed along our walk, even though he was in civilian clothes. He was easily recognizable due to his eye patch, acquired after he lost an eye during the firing of a sounding rocket in his early years of rocketry.

RIF (1992) – Thiokol "Reduction in Force"

November 1992 was a bottom in the Aerospace cycle. Thiokol management had a reduced budget, which they felt was needed to improve propellant processing. They said (did they actually believe it?) that they didn't see any more need for the type of problem solving that I did. Besides which, I was highly paid for a technocrat due to the raises and bonuses that I had received during my 16 years of saving Thiokol's butt. "And a new CEO (pharoh) arose who knew not Salita".

Consequently, it was proposed that I be laid off. I was told that there was a battle between Vice President Al McDonald and Program Manager Terry Boardman in support of retaining me, and managers Don Ketner and Suresh Kulkarni in favor of terminating me. I sent a letter to the new CEO Robert Lindstrom listing how I had made critical contributions to Thiokol's well-being every year since 1979, some of which I have documented in these memoirs. To no avail.

I was RIF-ed on November 12. Later that week I received a phone call from my mother in Jerusalem telling me that her bladder cancer had returned. That was a fun week! I booked a flight for myself and my daughter Karen to visit her for several weeks in Israel. Upon my return to Utah, Jim Kliegel told me that he wanted me to come work for him at TRW, but he had to work out some details yet. By February I still hadn't heard back from him.

In February 1993 I received another call telling me that my mother was moving to a hospice facility at the Hadassah Hospital on Mount Scopus. I spent a month sleeping at my parent's apartment, walking back and forth to the hospital every day (to save bus fare and get my daily exercise). Still no word from Kliegel. As you can imagine, it was a tough time.

In order to avoid feelings of depression after returning to the U.S., I focused my attention on coaching the mini-mite recreational ice hockey team my five-year-old son Joshua was on. I had a key to the rink, so I was able to set up additional practice time at off-hours for those who were available and interested. Sometimes only a few of the kids showed up, but it was a good opportunity for me and Joshua to spend time together playing hockey.

Eventually, Jim Kliegel notified me that he was ready to offer me a job reporting to him at TRW in San Bernardino, but he could only guarantee funding for 9 months. I didn't want to move myself and my wife and son to California, only to find myself unemployed 9 months later. So I asked if I could report to him, but work at the large TRW facility in nearby Clearfield, Utah. Electronic media had become powerful enough that we could interact frequently. I also said I would be willing to travel down to SB frequently for face to face interaction. He was willing, and we agreed that I should plan to come to SB several times a month, although as time passed, the frequency became less and less. I started work in Clearfield in June 1993.

Me

Joshua

Chapter 5 – Rocket Science with TRW (1993-2004)

The Problem: My first task under Jim Kliegel was to model slag generation, due to two pressing issues: (1) the flight failure of SICBM believed due to slag ejection, and (2) the high slag weight being generated in the Titan SRMU. The first step in generating a flowfield solution in a combustion chamber with submerged nozzle is to <u>generate a body-fitted multi-zone computational grid</u>. Unfortunately, I didn't have access to Thiokol's grid generator or their two-phase CFD code SHARP. Consequently, I had to provide my own grid generator and two-phase CFD code. However, without a plot package, you're working blind trying to create a computational grid or run a CFD code and interpret its predictions. So I first had to acquire or create a plot package for the computer platform on which the calculations were to be made. Initially that platform was a VAX, but after 1994 it was a PC.

(5A) <u>Development of an EXCEL Plot Package for Engineering Applications</u>

My Solution: Modeling of the two-phase flowfield in the rocket chamber requires a plot package for multi-zone body-fitted computational grids. As of the mid-1990s, no plot package was available that had these attributes for either mainframes or PCs. MATLAB for PCs and DI-3000 for VAXs could only plot flowfields on single-zone quasi-Cartesian grids. I got TRW to buy five inexpensive copies of the single-zone PC plot package from Dave Wilcox, but found that its treatment of the wall boundary conditions was erroneous. When I pointed this out to Wilcox, he decided that the effort to correct his package was greater than he wanted to make (he said that "he made more money selling his book on Turbulence Modeling"), so he took his plot package off the market.

Nonetheless, I needed a plot package for my chamber flowfield studies, ideally one that would work on both the VAX that I was using at TRW, and on my new Pentium PC at home during the evenings. I decided that the best approach would be

(1) to write a Fortran post-processor to convert the output from the CFD code into formats that needed to be plotted: scatter plots, computational grids, flowfield streamlines, velocity and direction vectors, contour lines, and color bands, then

(2) use the DI-3000 commands on the VAX and EXCEL commands on the PC to draw the lines and vectors.

This was fairly straight-forward on the VAX, but could EXCEL do the job for the PC? When I asked a Microsoft rep whether EXCEL could be used for engineering plotting, he said that Rockwell had asked the same question, and the answer was no. However, EXCEL seemed like the best option since it was free and portable for any computers with Microsoft Office. So:

1) I studied the Visual Basic manuals (which were still printed on paper), and was able to generate an **EXCEL Macro** (which I call **AUTOPLOT**) to draw the lines, arrows, contours, etc, that need to be plotted.

2) I converted my VAX processor over to the PC and modified it (**XLSETUP**) to write the CFD output to a spread-sheet in exactly the form required by AUTOPLOT. This was the key step in the process.

3) I created a second processor **MODIFY** to allow the user to merge multiple plot files, scale the variables (e.g. between metric and English units), filter the data, interchange dependent variable columns, etc.

4) I wrote a menu-driven DOS batch file **PLOTXL** to automate the entire process:

```
-------------------------------------------
BATCH FILE "PLOTXL" FOR "EXCEL" PLOTTING
-------------------------------------------
```

```
Choose option for plot file %INFILE%:

   0) NONE ... Terminate


   1) EDIT      plot file %INFILE%
   2) MODIFY    plot file (scale, filter, switch, sparse, contour, merge)

Usually:
   3) PLOT      resulting file using XLSETUP and EXCEL macro

Rarely:
   5) PLOT      EXCEL.DAT directly with standard EXCEL macro
   6)                          with colorized EXCEL macro
   7) Edit      file XLSETUP.INC to change array sizes
   8) Read      memo Salita_Plot_Packages
   9) BROWSE    User's Guide for XLSETUP
```

XLSETUP (processes plot files)

```
CHOOSE TYPE OF PLOT ...
-3) NONE ... TERMINATE
-2) CONTOUR PLOT W/UP TO 12 SYMBOLS REPLACING COLORS
-1) SCATTER PLOT (UP TO 14 CURVES Yn(X))
    DRAW COMPUTATIONAL DOMAIN ON 2D (I,J) MULTIZONE
    BODY-FITTED GRID
 0) DOMAIN BOUNDARIES ONLY
 1) BOUNDARY + J=CONSTANT LINES (also streamlines
    or O-ring shape)
 2) BOUNDARY + I=CONSTANT LINES
 3) GRID (COMPLETE)
    DRAW FLOWFIELD PLOTS
 4) CONTOUR LINES
 5) VELOCITY VECTORS
 6) PROPERTY PROFILES
 7) WATERFALL PLOT
 8) FLOWFIELD COLOR-BAND PLOTS
 9) CELL-PROPERTY COLOR  PLOTS
```

This EXCEL plot package (Ref 41) is very versatile (it generated the examples on the next two pages, as well as most of the plots in these Memoirs). A comparable macro **MLSETUP** for MATLAB was later constructed at Northrop Grumman by Kevin Zhou under my guidance, and I converted PLOTXL into a driver **PLOTML** for plotting with MATLAB.

Both plot packages are available to anyone who wants them by contacting me at mark_salita@yahoo.com.

Mach Contours and Velocity Vectors on a 5-Zone Grid (see grid on page 49)

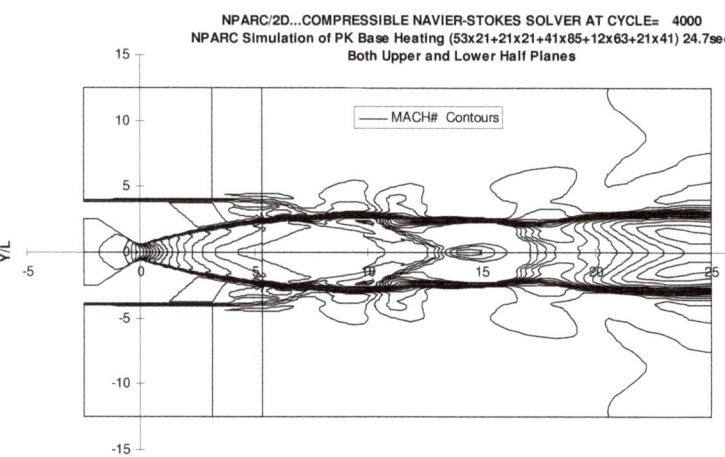

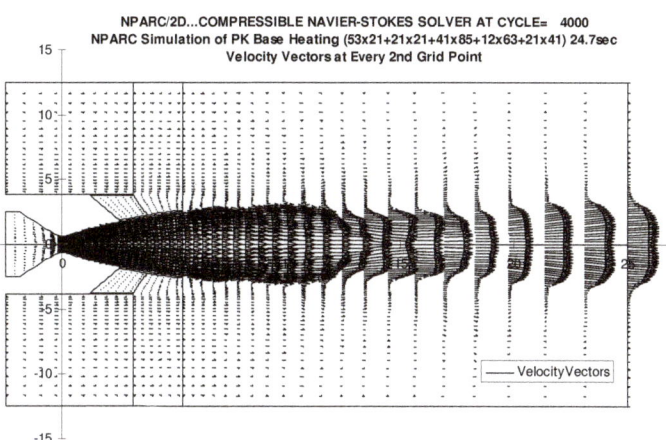

Plume Color-Band Plots with Superposed Contour Lines

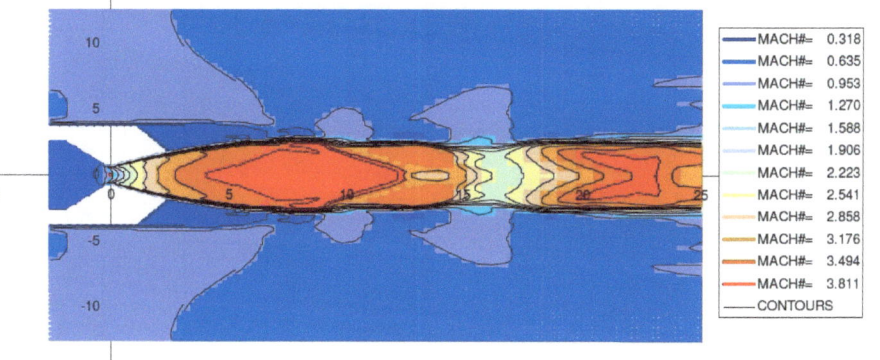

Extruded O-Ring

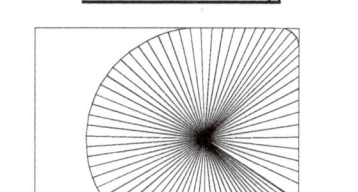

Star-48 Flowfield with Recirculation Zone

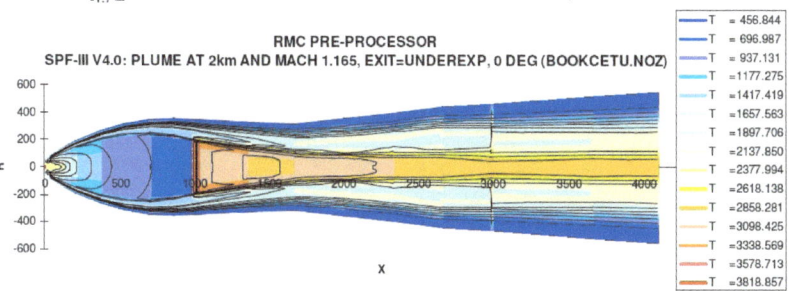

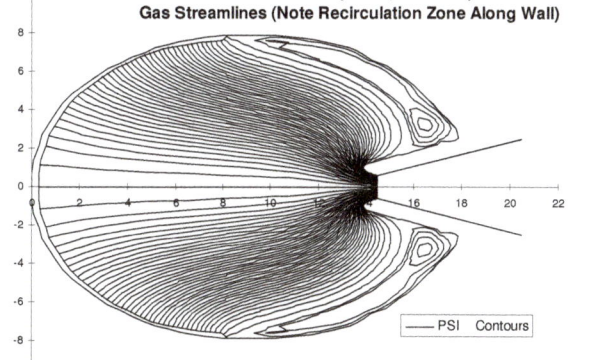

Scatter Plot of Wall Temperature versus Location and Time

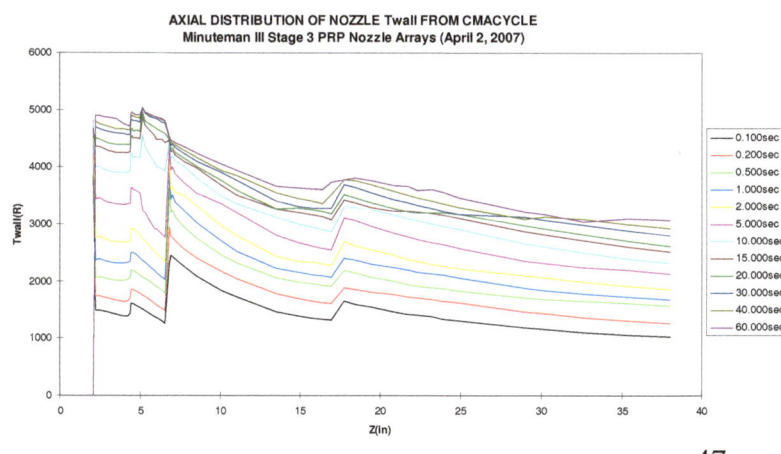

Scatter Plot of Absorption Band-Model for H$_2$O

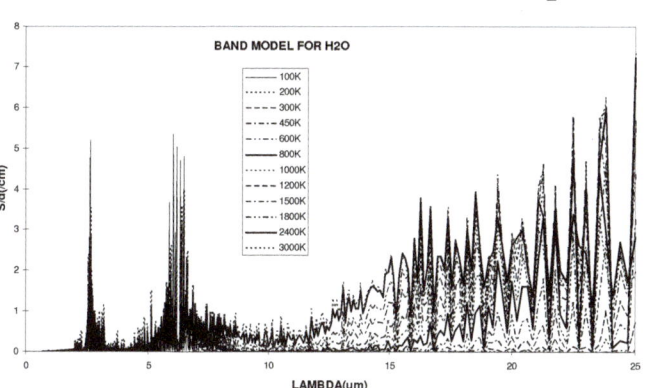

Plume in Rarefied Atmosphere Predicted by Russian Code NARJ

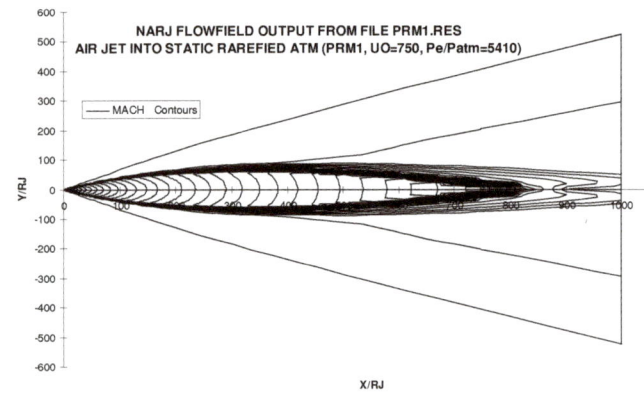

Waterfall Plot of Pressure versus Time and Axial Location During Shuttle Booster Ignition

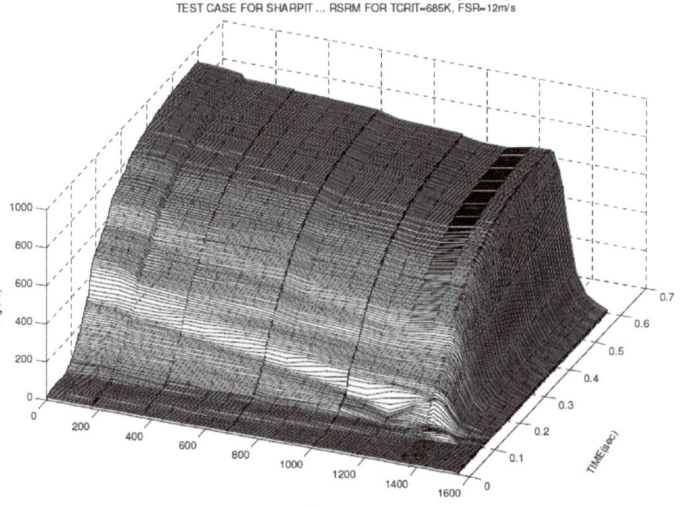

Nozzle Flowfield

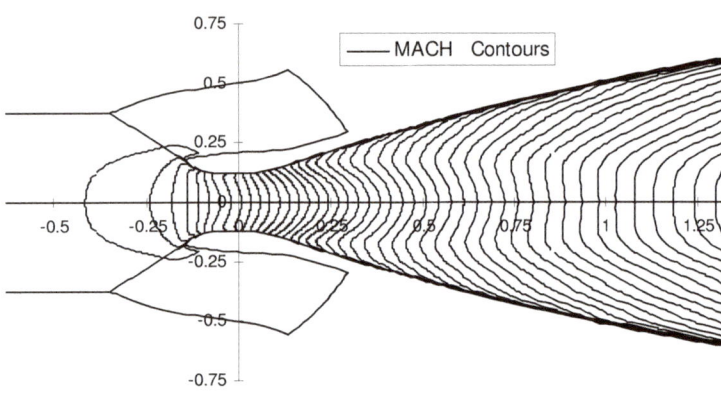

Nozzle Material by Color

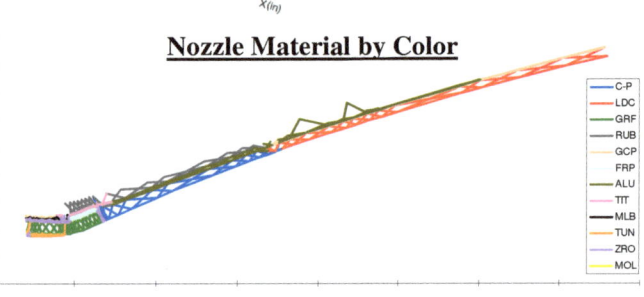

Gas Direction Vectors in Combustion Chamber

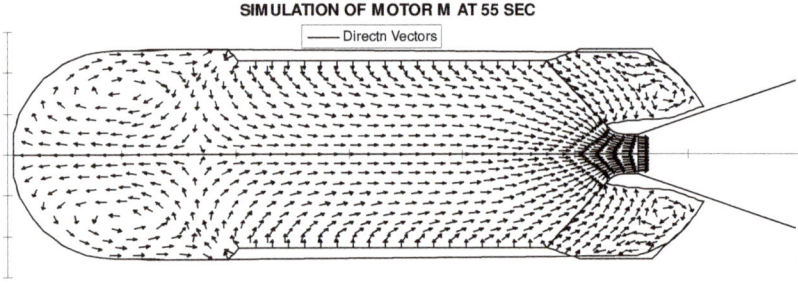

Mach Contours in Rarefied Plume Flow

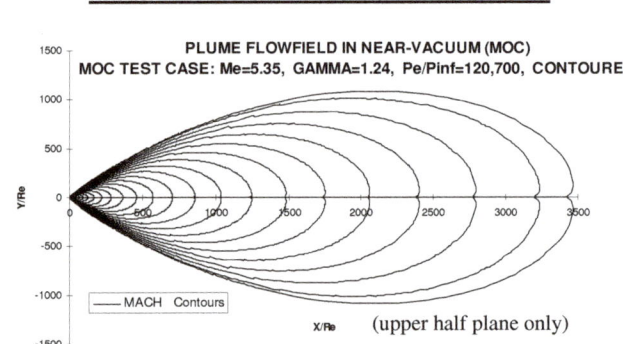

World Map with Superposed Missile Trajectory
(North Korea to LA)

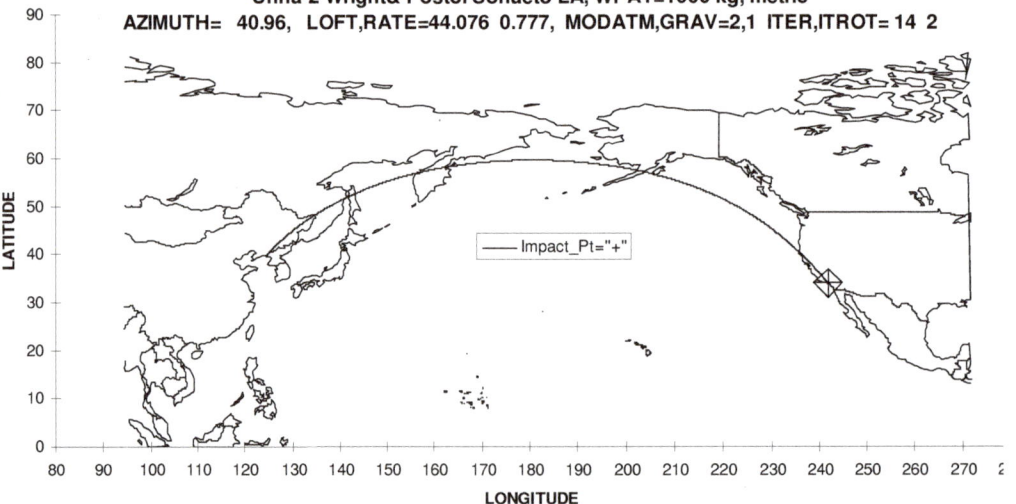

48

(5B) Development of CFD Chamber Flow Modeling Package EVT (1994)

The next step was to create an axisymmetric multi-zone CFD package **EVT** comprised of (1) a body-fitted computational-grid generator (originally my elliptic grid generator **EGG**), (2) a Navier-Stokes viscous gas flowfield solver **VORSTREM** formulated in **vor**ticity/**stream**function dependent variables, and (3) a Lagrangian droplet tracker **TD2P**. EVT turned out to be extremely robust, versatile, useful, and easy to use as will be demonstrated throughout these Memoirs.

Grid Generation (1993)

The Problem: TRW/San Bernardino had a site license for the commercial grid generator GRIDGEN but wouldn't extend it to cover me as their only user in Utah. Since I also didn't have access to the Thiokol grid generator, I had to write my own. The focus of slag accumulation is on the aft end of a combustion chamber with submerged nozzle. Creating a computational grid that would be "body-fitted" to that kind of configuration was generally not available in the industry at that time. Such body-fitted gridding was only recently becoming available in commercial CFD packages. Golafshani and Loh had recently created an elliptic grid generator at Thiokol patterned after the code GRAPE developed at Mississippi State.

My Solution: Since I didn't have access to the Thiokol code, I created my own axisymmetric Elliptic Grid Generator (**EGG**, Ref 42). However, elliptic grid generators are not very useful for computational domains requiring multiple zones. So I also created a Multizone Algebraic Grid Generator (**MAGG**, Ref 43), which allows for an arbitrary number of axisymmetric contiguous zones. Examples are shown below for a combustion chamber (only the upper half-plane is shown and needed for axisymmetric configurations); these grids stop at the nozzle throat because the incompressible flow solver VORSTREM described on the next page only extends to the nozzle throat. EGG worked well only at later burn times, while MAGG worked well at all but the late burn times:

MAGG 2-Zone Grid (at 20 sec)   EGG 1-Zone C-Grid (at 54 sec)

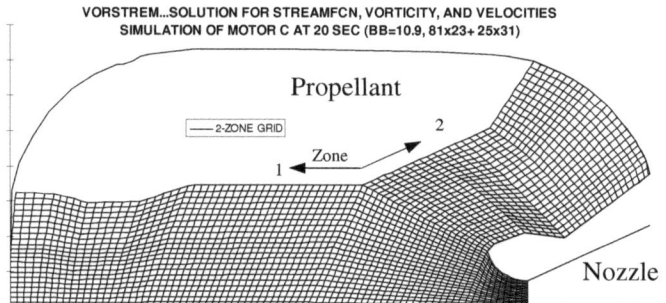 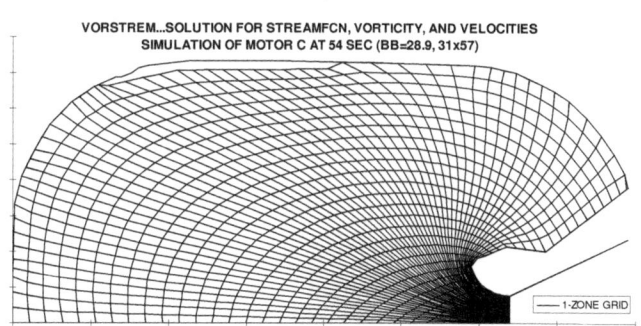

MAGG was also useful for domains beyond the combustion chamber. A 5-zone MAGG grid is shown below for modeling the nozzle and plume flowfields for Peacekeeper during flight using the CFD code NPARC (see 3 plots at top of page 47):

Boundaries of the 5 Zones   Grid for the 5 Zones

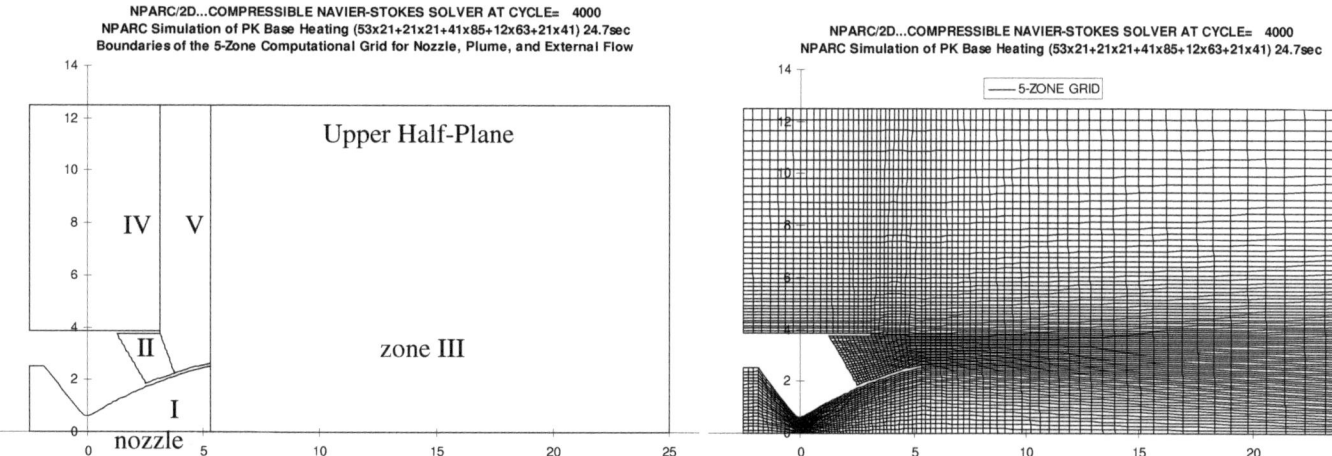

MAGG was also useful for 3D flows past axisymmetric vehicles. By rotating an axisymmetric grid about its axis and storing the coordinates at circumferential angular increments, a grid for computation of the 3D flow around an axisymmetric vehicle flying at angle of attack could be generated. This was accomplished by creating a code **ROTATE**, which I used frequently to generate the 3D grids when running the commercial CFD codes like CFD-ACE.

VORSTREM – Simple But Approximate CFD Code to Model Gaseous Flow in Rocket Chambers

The Problem: By the late 1980's, numerous compressible CFD codes (CELMINT, SHARP, PHOENIX) were available to model the gas flow in an SRM chamber. However, setting up the input, running the codes, and processing the output was a very time-consuming process. A simplified method was needed. The principal simplified method had been to approximate the gaseous chamber flow as axisymmetric and "potential" (i.e. <u>irrotational and inviscid</u>). The flow in many rocket chambers is indeed axisymmetric, and propellant geometries that are initially three-dimensional usually burn back to quasi-axisymmetric relatively early in the combustion process. Unfortunately, the real chamber gas flow is very <u>rotational</u>, and the flow behind submerged nozzles is <u>highly viscous</u>, so potential flow is very inaccurate. A more-appropriate model was needed.

My Solution: I realized that the gaseous flow in the rocket chamber was basically incompressible (Mach numbers rarely exceed 0.3). I had had good success modeling incompressible propellant flow during mixing and casting using the (viscous) Navier-Stokes equations written with vorticity ω and streamfunction Ψ as dependent variables, instead of primitive-variables (velocity components u and v, density ρ, temperature T). Indeed, the resulting equation for streamfunction is identical to that for potential flow, except for a source term that accounts for flow rotation in the form of vorticity ω. The vorticity is determined by a second equation in the same form as the streamfunction equation. Hence, the resulting equations (11,12 below) are only twice as expensive to solve, but account fully for flow rotationality and viscosity. Using streamfunction Ψ as one of the dependent variables has the added benefit that a line of constant Ψ represents a gas streamline.

The solution of axisymmetric flowfields in irregular configurations is simplified by transforming the physical domain (x,y) to a rectangular computational domain (ξ,η) whose grid will be uniformly spaced. The resulting flow equations in transformed coordinates and normalized time $\tau=Ut/L$ then become (with subscripts signifying partial derivatives):

$$D\Psi_{\xi\xi} + 4M\Psi_{\xi\eta} + E\Psi_{\eta\eta} + 2C^{-}\Psi_{\xi} + 2H^{-}\Psi_{\eta} = y^{\varepsilon}\omega \qquad (11)$$

$$D\omega_{\xi\xi} + 4M\omega_{\xi\eta} + E\omega_{\eta\eta} + 2C^{+}\omega_{\xi} + 2H^{+}\omega_{\eta} = [G+\omega_{\tau}]\frac{Re}{\mu} + \frac{\varepsilon\omega}{y^{2}} - S$$

where $\qquad (12)$

Transformation of physical grid to rectangular space.

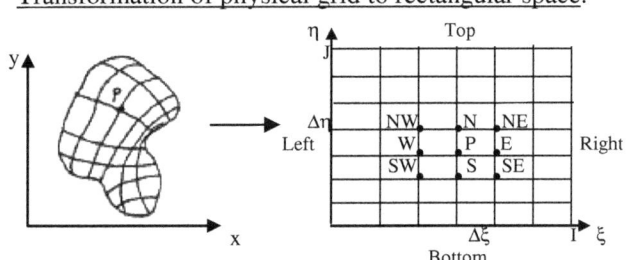

$$D = \xi_{x}^{2} + \xi_{y}^{2} \qquad\qquad C^{\pm} = \tfrac{1}{2}(\xi_{xx} + \xi_{yy} \pm \varepsilon\xi_{y}/y)$$
$$E = \eta_{x}^{2} + \eta_{y}^{2} \qquad\qquad H^{\pm} = \tfrac{1}{2}(\eta_{xx} + \eta_{yy} \pm \varepsilon\eta_{y}/y)$$
$$M = \tfrac{1}{2}(\xi_{x}\eta_{x}+\xi_{y}\eta_{y}) \qquad G = \xi_{x}(\omega u)_{\xi} + \eta_{x}(\omega u)_{\eta}$$
$$S = \text{turbulence terms} \qquad\quad + \xi_{y}(\omega v)_{\xi} + \eta_{y}(\omega v)_{\eta}$$

The solution of these equations at a single time τ was automated as program VORSTREM (Ref 44) and allows for the equations to be solved in several separate but contiguous zones, with appropriate continuity at their interfaces. The equations are time-marched from an initial guess until convergence to steady state. Laminar viscosity was initially assumed (S=0), with the intent to add turbulence. However, the incompressible laminar predictions in the chamber were subsequently shown to agree quite well with those from fully-compressible turbulent codes; consequently, no turbulence model was ever added. The gas streamline pattern and direction vectors are plotted below for the motors whose grids were shown on the previous page.

I am very proud of this code; it is simple (only 1624 lines of code), it is elegant in its use of classical equations on multi-zone body-fitted grids in transformed coordinates, it is very robust (e.g., see pages 52, 55, 75), and runs in seconds on a PC.

Gas Streamlines

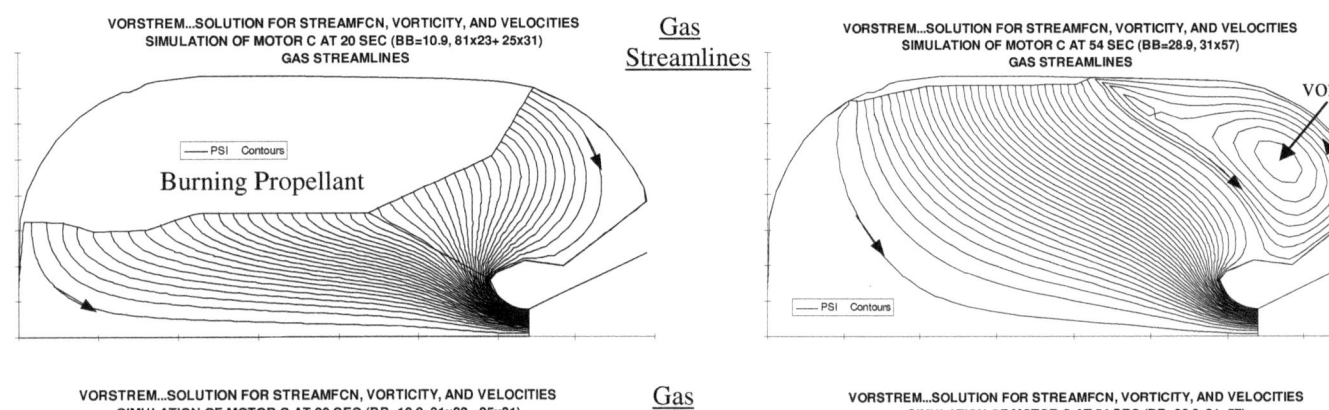

Gas Direction Vectors

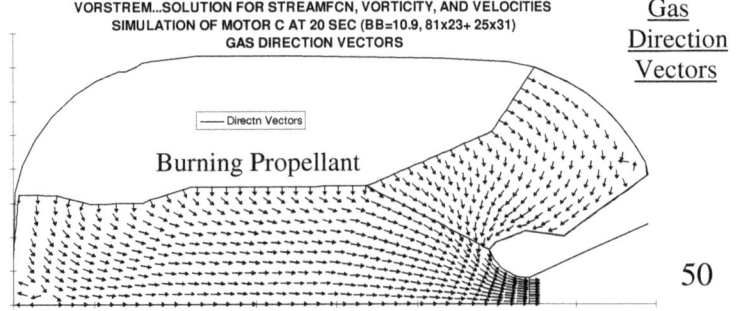

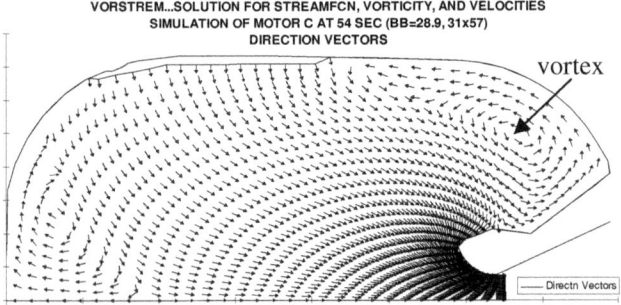

TD2P – A Lagrangian Droplet Tracker

Most solid rocket propellants contain 5-25% aluminum powder as fuel. Quench Bomb testing discussed earlier showed that when the propellant burns, the aluminum melts and oxidizes to aluminum oxide (Al_2O_3) droplets, with 60-90% in the form of smoke of mass-median diameter 1.5μm, and the remainder as "coarse" droplets of mass-median diameter 20-150μm. These droplets can comprise up to 35% of the mass fraction of the combustion products of the propellant. The use of aluminum greatly increases the thrust of the motor and suppresses pressure oscillations, but the formation of the Al_2O_3 causes a number of detrimental effects including (1) some loss of thrust due to their velocity and thermal lags, (2) impingement erosion in the rocket chamber and nozzle, and (3) droplet entrapment as "slag" in the submergence cavity behind submerged nozzles.

The Problem: The accumulation of Al_2O_3 slag behind the submerged nozzles in SRMs became important in the late 80's and early 90's due to a number of motor anomalies. The excessive slag weight in Shuttle Booster ETM-1A was discussed earlier. On the other hand, subsequent static tests of Shuttle boosters using the new WECCO ammonium perchlorate (AP) oxidizer were generating slag weights that were only a third of that generated using Kerr-McGee AP. This large variability had to be understood. Furthermore:

(1) The first flight test of the SICBM missile failed when the slag pool in the first stage was ejected asymmetrically out the nozzle during tailoff, which occurred at the same time as 1/2 staging. This caused a nozzle side force on the first stage nozzle that swung the first stage head-end such that it damaged the stage 2 nozzle (see page 44).
(2) Slag weights measured after static tests of both the Titan SRMU and Ariane5 revealed over 4000 lbm. These slag weights were huge and of concern.

My Solution: The droplets are injected at the propellant surface into the gas flowfield generated by VORSTREM and tracked with a Lagrangian method (follow each droplet) using my droplet tracking program **TD2P** (Two-Dimensional Two-Phase, Ref 45). Care had to be taken to treat properly the possible reverse droplet flow in the recirculating region. An example is shown below for droplets of diameter 75.8μm assuming vehicle axial acceleration of 10 g's at 20 sec and 0 g's at 54 sec:

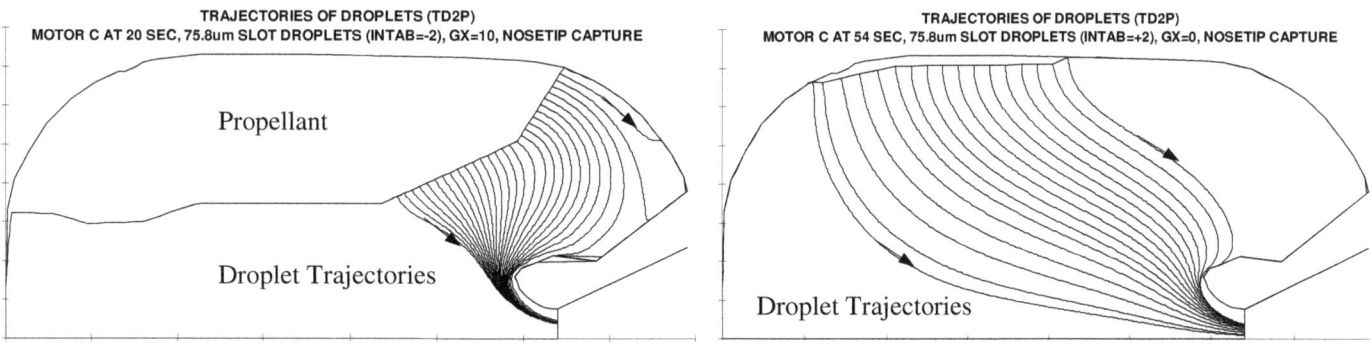

The original TD2P model assumed inert droplets, i.e. the droplet was everywhere liquid with constant diameter. However, the "Lead Pellet Problem" discussed in Chapter 6 required the addition of a melting/vaporization option to shrink the droplet.

Making EVT User-Friendly

EGG, VORSTREM, and TD2P were coupled together as a package EVT (Ref 45). EGG (or often MAGG) was used first to generate a computational grid, which was read by VORSTREM, which generated a flowfield file which was read by TD2P, which solved the droplet trajectories. This sequence was automated by creating a menu-driven DOS batch file RUNEVT, which is so efficient that the Air Force requested its use numerous times for critical anomaly investigations (see pp 53, 75).

```
WHICH CODE DO YOU WANT TO RUN?               VORSTREM...MAKE CHOICE FOR "%INFILE%"
    0) NONE ... QUIT                             0) RETURN TO MAIN MENU
    1) VORSTREM to simulate gas flowfield        1) EDIT VORSTREM INPUT FILE = "VORSTREM.%INFILE%"
    2) TD2P      to track droplets               2) RUN  VORSTREM
    3) HEATSPH  to calculate heating of droplets 3) EDIT OUTPUT FILE "VORSTREM.OUT"
                                                 4) EDIT/PLOT   FILE "VORSTREM.STO"
                                                 5) PLOT FLOWFIELD (USING MATLAB OR SALITA'S EXCEL MACRO)
  TD2P...MAKE CHOICE FOR "%INFILE%"                   (WILL ADD CASE BOUNDARY IF YOU HAVE FILE %INFILE%.WAL)
    0) RETURN TO MAIN MENU                       6) PLOT FLOWFIELD + CASE BOUNDARY
    1) EDIT TD2P INPUT   FILE = "TD2P.%INFILE%"  9) BROWSE VORSTREM USER'S GUIDE
    2) RUN  TD2P
    3) EDIT TD2P OUTPUT  FILE = "TD2P.OUT"
    4) EDIT TD2P PLOT    FILE = "TD2P.STO"
    5) EDIT TD2P NUSSELT FILE = "TD2P.NUS"
    6) EDIT TD2P MELTVAP FILE = "TD2P.SUM"
    7) EDIT TD2P CONCENT FILE = "TD2P.FLO"
    7) ADD CASE BOUNDARY FROM SPP
    7) RUN  INFLOW (TO GENERATE INDEPENDENT INFLOW GAS FLOW RATES)
    8) PLOT FLOWFIELD + CASE BOUNDARY FROM "DRAWCASE.PLT"
    8) PLOT TRAJECTORIES (IFF YOU HAVE SALITA'S XLSETUP AND EXCEL MACRO)
          (WILL ADD CASE BOUNDARY IF YOU HAVE FILE %INFILE%.WAL)
    9) BROWSE TD2P USER'S GUIDE
```

National Propulsion Aerodynamic Research Code (NPARC): Failed for Rocket Internal Flow

NPARC originated as the Aerodynamic Research Code (ARC), then was specialized for propulsion applications (PARC), and finally was provided free to any U.S. organization, with support provided by NASA Glenn and AEDC. The code has many attributes (e.g. multi-zone, body-fitted grids), and many limitations (e.g. non-reacting, perfect gas). It works well for external flow modeling; however, when I tried using it for application to solid rocket internal problems, I soon discovered fatal flaws.

One flaw was a requirement that the computational grid be orthogonal (perpendicular) to the domain boundaries. That was too limiting for most grid generators, so AEDC helped me generalize the coding to allow non-orthogonal boundary grids.

However, more importantly, the code was not "pre-conditioned" to allow stable calculation of flows with a wide range of Mach number (e.g., chamber to nozzle exit). Consequently, when I tried to model the flowfield inside the Star48 Space Motor and nozzle, the solution was totally erroneous on both of two different types of computational grids (C-grids and O-grids):

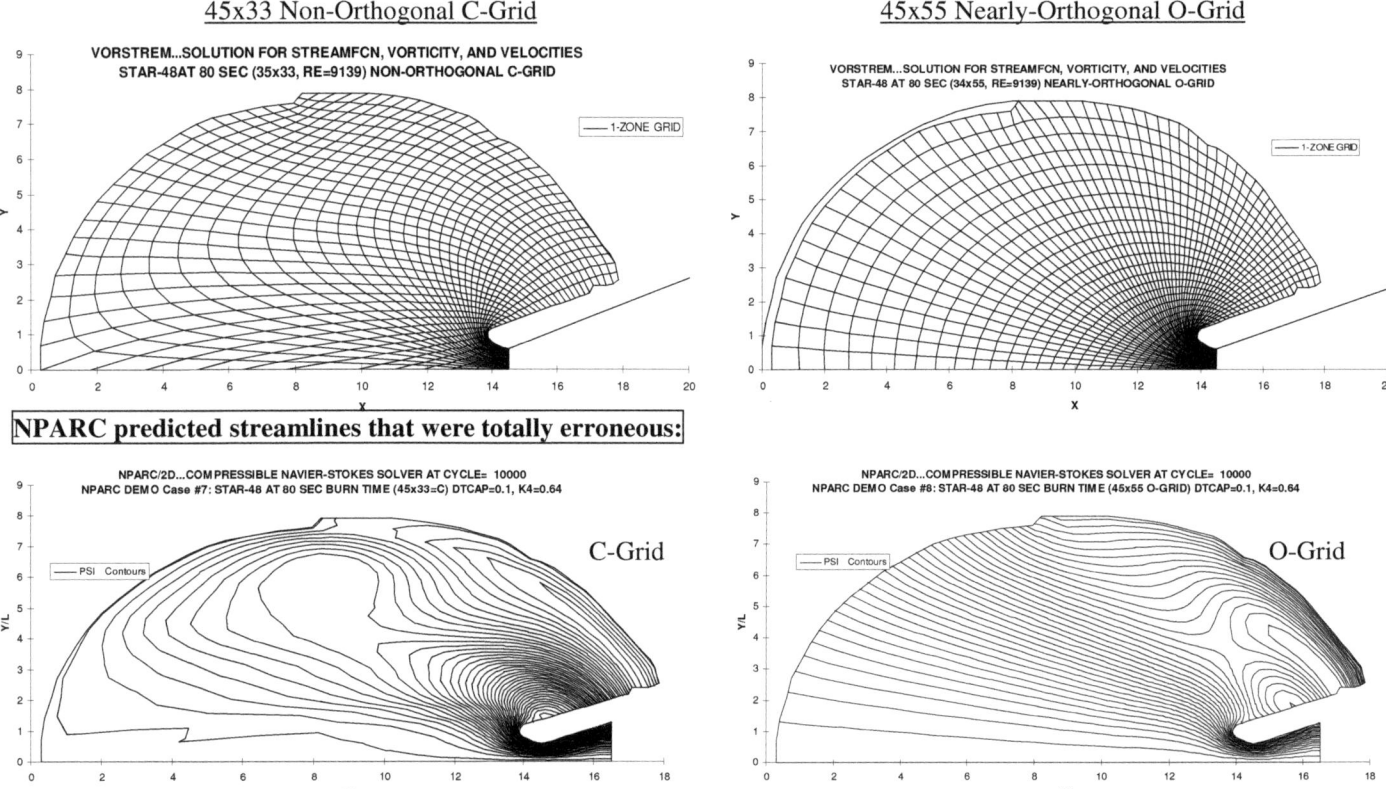

NPARC predicted streamlines that were totally erroneous:

By contrast, the solutions predicted by **VORSTREM** were well-behaved and nearly identical, as required for different grids (a nearly-orthogonal O-grid, and both nominal and quadrupled highly non-orthogonal C-grids), demonstrating its robustness:

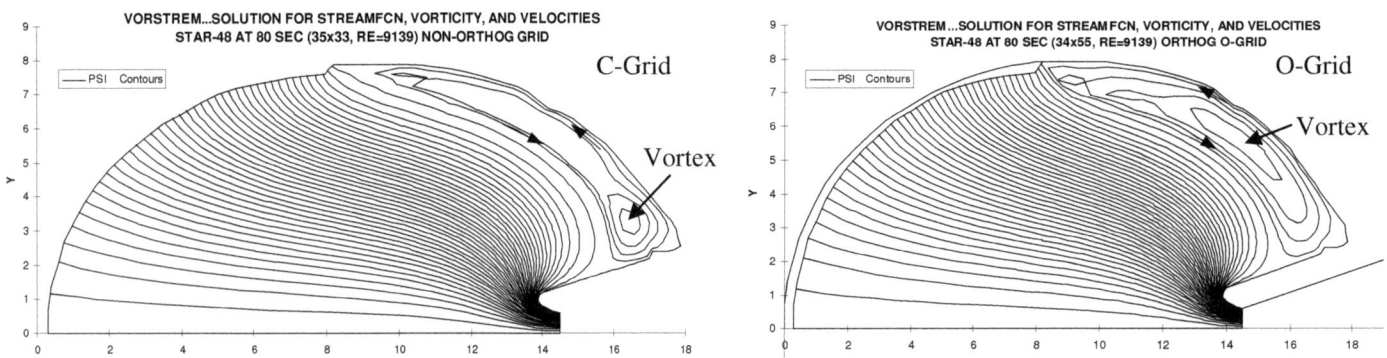

The NPARC code was dropped soon after by NASA/AEDC and replaced by the WIND-US code. My discoveries of NPARC flaws may well have been a factor in that decision, as well as the need to include non-perfect reacting gas and other features.

I also ran the commercial compressible CFD code **CFD-ACE** on both O- and C-grids. The solution on the nearly-orthogonal O-grid agreed closely with that from VORSTREM. However, the CFD-ACE solution on the non-orthogonal C-grid did not expose the recirculation zone until the grid-point density was quadrupled; only then was the solution nearly identical to that predicted by VORSTREM on the sparser grid. This again emphasized the robustness and efficiency of VORSTREM.

RIF-ed Again, But Rehired Due to EVT (1994)

Even though my TRW office was in Ogden, my funding and technical supervision (Director Jim Kliegel and immediate supervisor Bernie Morton) came from San Bernardino. However, there was a company rule in 1992-94 that I had to have an administrative supervisor at my office location (Ogden). Consequently, I reported to him and did support some of his tasks, even though he felt that his management of me was more a burden than a benefit.

Unfortunately, after only one year of working for TRW, the company lost a major Air Force contract that had funded about 30% of the engineers in the division. The supervisors were forced to decide who to keep and who to RIF. Since I was an unnecessary appendage for my Ogden supervisor, he chose to add me to the RIF in March of 1994. This occurred in spite of the fact that I had already written an AIAA paper about VORSTREM that I had committed to present at the 27th Joint Propulsion Conference in Indianapolis three months later. In addition, based on my recent slag modeling with EVT (Ref 46), I had already been selected to Chair a JANNAF Workshop on Slag Generation to be held in conjunction with the JPC.

What to do? TRW would not pay for my trip to Indianapolis since I was now an inactive (but recallable) employee. I decided that I should go to Indianapolis anyway due to my commitments. Maybe I could get wind of a new job. My supervisor Bernie Morton generously offered to pay for the cost to print the required 100 copies of my paper then required by AIAA. But I would have to pay for the entire cost of the trip myself. Not a very happy option considering I was now unemployed.

Surprisingly, as I will now recount, I was able to undertake the full 4-day trip at a cost of only $200. Here's how:

1) Luckily, there was an airfare war on, and I was able to get round-trip tickets to Indianapolis for only $150.
2) My wife Alice drove me to the Salt Lake City airport on Saturday.
3) I had brought enough portable food on the plane to take care of dinner Saturday night.
4) I caught a free hotel shuttle bus between the Indianapolis airport and my hotel, within walking distance of the Convention Center that also included free breakfast.
5) I had arranged with my colleague Doug Coats to share a hotel room, and he offered to pay for my half of the room if I would man his sales booth at the Convention while he was marketing his software products or taking breaks.
6) I chaired the Slag Workshop on Sunday.
7) The fee for registering at the conference as unemployed was only $50, but this enabled me to attend the Sunday night gala sponsored by Allison Engines where there was plenty of great food. I took some back to the hotel with me for subsequent lunches.
8) I had free dinners by visiting Hospitality Suites sponsored by various aerospace companies at the conference.
9) On my return flight to Utah, I met several Thiokol employees who had attended the conference; one of them gave me a ride to Ogden and dropped me near my home, where my wife picked me up.

As amazing and ironic as it sounds, my effort still to chair the JANNAF Slag Workshop (Ref 47) got my job back at TRW!!! Here's how:

Attendance at the workshop required everyone to get pre-approval from JANNAF (which I received before my RIF). As the workshop began, Bob Geisler from AFRL happened to be walking down the hallway past the workshop room. He hadn't heard about the slag workshop, but he immediately asked to attend. Bob was so well known and respected in the JANNAF world that the screeners at the door let him attend without pre-approval.

Ironically again, during the duration of the conference, the first flight of the Extended Pegasus rocket failed in flight. Film of the failure showed a lot of slag being ejected, but it was unknown whether the slag ejection caused the failure or the failure caused the slag to eject. After the conference, Bob Geisler was asked by the Air Force to head a team to investigate the possibility that the slag had caused the flight failure.

Based on my presentations and chairmanship, Bob was aware that I was one of the slag experts in the industry, and the only one who had the tools (EVT) to conduct rapid modeling. So he wanted me to support his team. The easiest way to fund my support was to pay TRW to rehire me. My former San Bernardino supervisors were delighted, and I was rehired for the several months that I needed to generate the desired calculations (see pages 54, 55, 99). It turned out that an error in the software used for Guidance and Control was the cause of the flight failure, and not the slag!

After completion of my failure investigations, Morton and Kliegel found enough funding to keep me on staff, and got approval for me to report to them directly. I was to work for them from Ogden for the next 15 years until my voluntary retirement in 2009. They were well aware that due to my prolific documentation habit, they knew better what I was doing than what most of their other engineers in San Bernardino were doing.

This was the second occasion where a chance intersection of my path with someone else in a hallway was to have a major effect on my career (the first being the photo technician at MSFC). Lightning can strike twice!

(5C) <u>More Slag Generation and Modeling</u>

<u>The Problem</u>: There was a lot of misunderstanding about slag generation at that time (1994). For example, a paper was published in 1991 by an engineer at McDonnell Douglas that predicted the accumulation of slag in spinning space motors. His model generated slag weights that agreed with measured weights in spin tests. However, I noticed that

(1) the droplet diameters that he used in his model were an order of magnitude too small (he had erroneously used Hermsen's correlation for diameters that occur <u>after</u> they shatter in passage through the nozzle, rather than the appropriate diameters for the chamber <u>before</u> shattering), and

(2) he had inappropriately (as discussed above) assumed potential flow to predict the flowfield.

I called him on the telephone to explain that his erroneous flow model appeared to have counter-balanced the inappropriate choice of droplet diameters. His response was **"Don't bother me with science, my slag correlation matches the data."**

That paper was not the only slag paper masquerading as correct. I was a reviewer for an article submitted to the Journal of Propulsion and Power that was bad enough that I (and at least one other reviewer) recommended that it not be published. Lo and behold, some months later the article appeared in the Journal of Spacecraft and Rockets. I called the editor of JSR and asked him if he knew that the article had been shot down by JPP on technical grounds. He did not know that. So remember that **there are refereed journal articles that should never have been published**.

<u>My Solution</u>: I wrote an article that was published in the Journal of Propulsion and Power entitled "Requirements and Deficiencies in Modeling of Slag Generation in Solid Rocket Motors" in February 1995 (Ref 4). Previous modeling of slag accumulation had assumed that all droplets were coarse, which therefore required that slag deposition begin late in the motor burn in order to generate the correct measured slag weights. However, the presence of so much previously-unknown smoke (which is too small to be trapped as slag) meant that the retention of fewer coarse droplets required that slag accumulation had to begin early in the motor burn. The concept of an early start to slag accumulation was subsequently confirmed by Real-Time X-Ray Radiography video that could see through the cases during the vertical static firings of SICBM and Titan SRMU motors. The history of slag volume was deduced digitally from the RTR by Bob Fredrick for SICBM (shown below) and by Catherine Dolan for SRMU (similar looking, see Ref 4), and clearly showed the early onset of accumulation.

> However, I extracted much more from this RTR, as documented in Ref 4, which led to a much better understanding of the slag accumulation process throughout the industry. For example:

(1) a plateau in slag volume is reached when the slag pool behind the submerged nozzle becomes "aerodynamically full", and

(2) the boiling of the slag pool that occurs during rapid motor depressurization at the end of burn ("tailoff") causes the slag pool to expand greatly, but not enough to boil over (at sea level pressures) and expel significant slag mass. Consequently, the slag measured after the sea level static firing is indeed nearly that generated during the firing (little was ejected).

The RTR showed identical qualitative behavior of the slag pools for both the SICBM and SRMU static firings, which occurred at sea level (1.0 atm). However, years later, Mulrooney (Ref 48) showed large slag ejection during tailoff of several SRMs in flight. He attributed this to the fact that the vehicle was at high altitude during tailoff, so the motor and atmospheric pressure was much lower than for ground testing, so the boiling process was much stronger and longer in duration.

Real-Time X-Ray Radiography Data Showing History of Slag Pool Volume in Vertically-Fired SICBM Motor.

A very similar plot for an SRMU static firing was generated (see Ref 4).

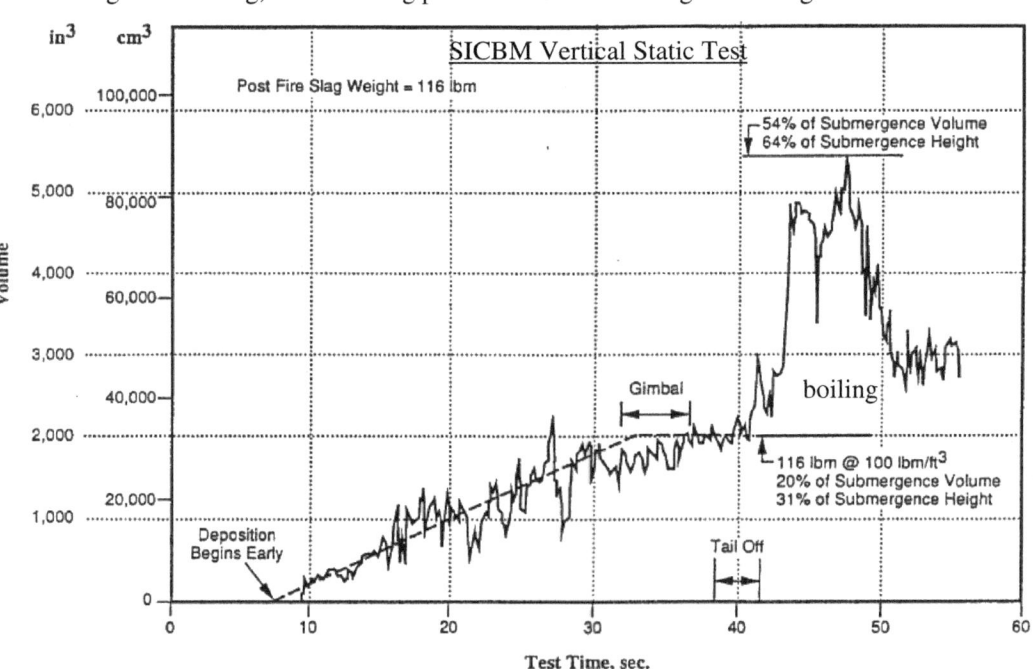

More Evidence of EVT Robustness: (1) Droplet trajectories predicted by EVT were in good agreement with those from compressible turbulent CFD codes. For example, note the close comparison to NASA-sponsored CFD code FDNS (Finite-Difference Navier-Stokes) for 100μm droplets in the aft end of the Shuttle Booster at 80 sec burn time:

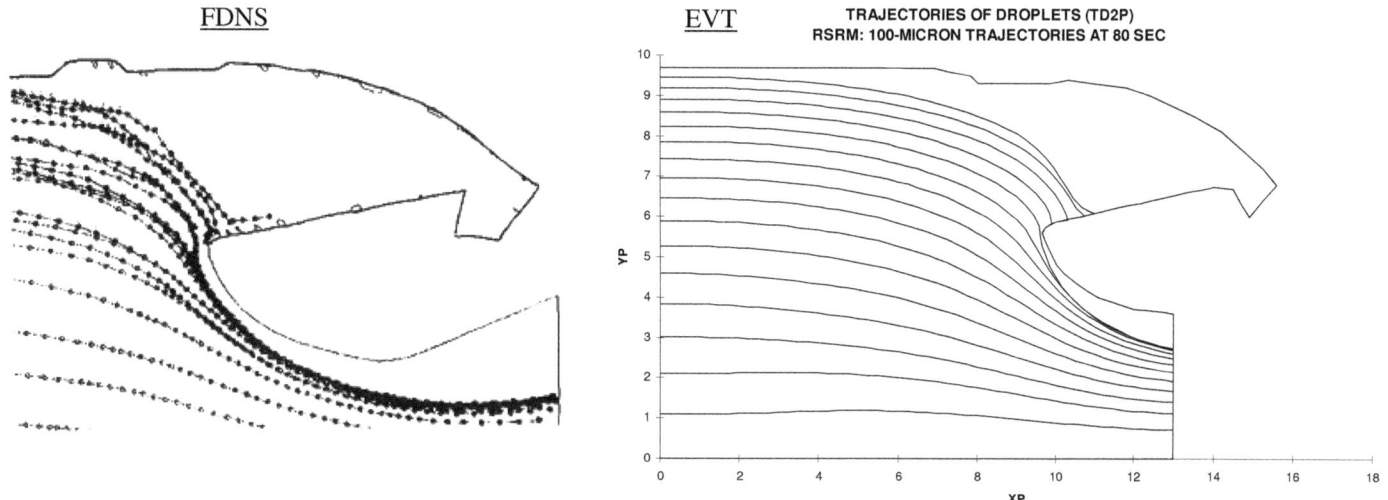

(2) EVT was able to explain the post-fire slag residue in a vertically-fired Titan SRMU motor: the slag crust found along the outer wall was due to the "slinging" effect of the vertical flow in the recirculation zone. (3) EVT showed that the presence of the slag pool didn't affect the core flow, just squeezed (and sped up) the recirculation vortex; hence the aerodynamic plateau.

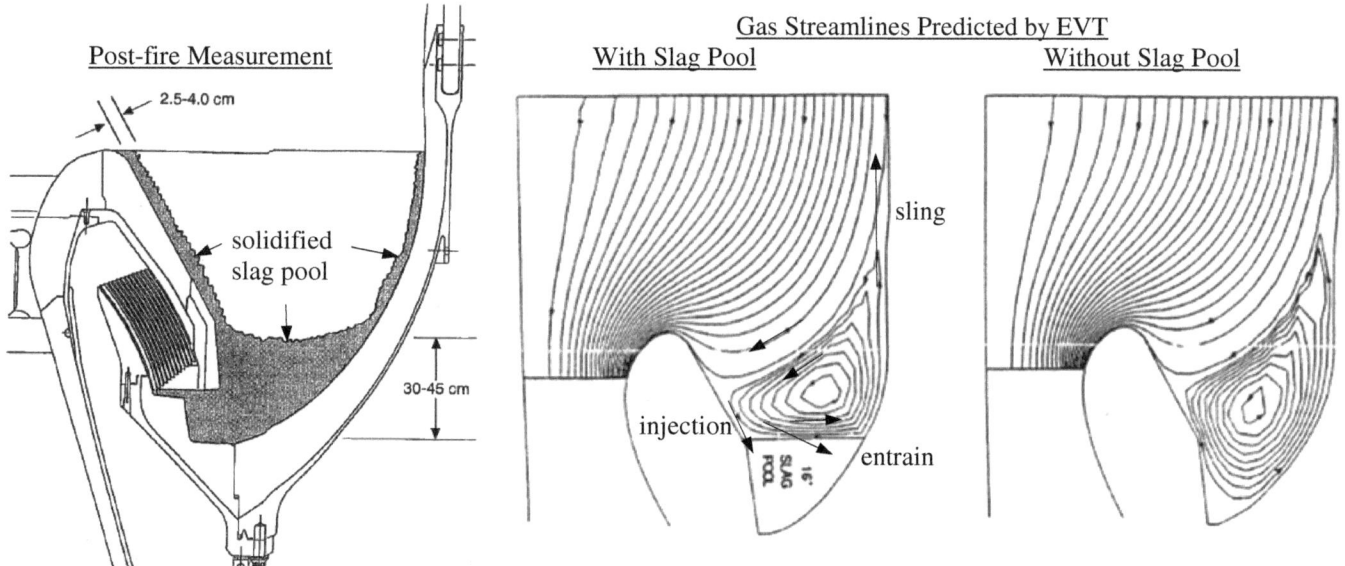

(4) EVT allowed predictions to be made rapidly for eight different motors (Ref 45). Predicted slag weights were calculated for three different slag capture rules: (1)were droplets captured only by entrainment in the recirculation zone (see figure on page 42), or (2) also by impingement aft of the submerged nosetip, or (3) by some other criterion?:

Motor	Measured	Nosetip	Slope Break	Recirculation	
			Slag Weight (lbm)		
			Predicted by EVT		
SICBM	128	202	158	10	
SRMU	4366	4842	2931	594	
RSRM	500-3500	5424	3170	2085	(see average on page 42)
Castor IVb	12	463	263	161	(no explanation for overprediction)
Pegasus XL	40	520	---	120	
Pegasus STD	n/a	490	---	120	
Star-48 (rpm=0)	0	43	---	0.3	
Star-37 (rpm=0)	1.2	---	---	4.9	

From these results I concluded that (1) no single capture rule was correct for all motors (at least based on predictions using EVT), and (2) the extremely low slag measured for Castor IVb was due to an unknown advantage provided by the grain design, since quench bomb data showed that the droplet size distribution was comparable to other AP-composite propellants.

(5D) Motor Ignition – More Issues

Ignition Overpressure (IOP) During Launch of ICBM from Minuteman Silo (1997)

Remember how ignition of the Shuttle booster caused a steep-fronted pressure wave to exit the nozzle and reflect from the exhaust hole in the MLP (page 25). Similarly, when the first stage of a Minuteman ICBM ignites during launch from a silo, the compression wave exits the nozzle and reflects from the silo floor. Whereas the pressure wave during the Shuttle launch can expand and weaken in the open air of the launch pad, the pressure wave in the silo is confined by the silo walls, so it remains strong as it passes over the missile (a concern). In addition, the wave reflects back into the silo from its open top as an expansion wave, thereby weakening the upward-moving compression wave. Superposition of the upward compression wave and the downward expansion wave at each location in the silo results in a triangular overpressure history, whose shape depends on its distance from the floor and the silo opening (see figure below for silo station 805 near the missile nozzle).

The Problem: There are two types of silos: shallow and deep, where the deep silo has 10 ft of additional distance between the nozzle and the floor. I was asked to explain why the peak overpressure (like a sonic boom) caused by motor ignition in deep silos is consistently about 50% greater than in shallow silos.

My Solution: Broadwell and Tsu had previously shown that the IOP in launches from Titan silos increased significantly due to afterburning of the fuel-rich rocket exhaust with the ambient air in the silo. So I postulated that the extra 10 ft and extra resulting volume below the nozzle in deep silos allowed for more space for afterburning with the larger volume of air in the deep silo than in the shallow silo. I then conducted a "screening" study (Ref 52) to determine the dominant chemical reactions in the afterburning process, and the forward rates K_f of these reactions. I generated a computational grid around the missile in the silo, and along with the reaction set ran simulations using commercial code CFD-ACE; colleague Les Glatt also ran his reacting CFD code NEQ, with similar results. Simulations were also run for $K_f=0$ to suppress afterburning. The results shown below at station 805, as well as those at the other measurement stations, strongly supported my postulation (Ref 53).

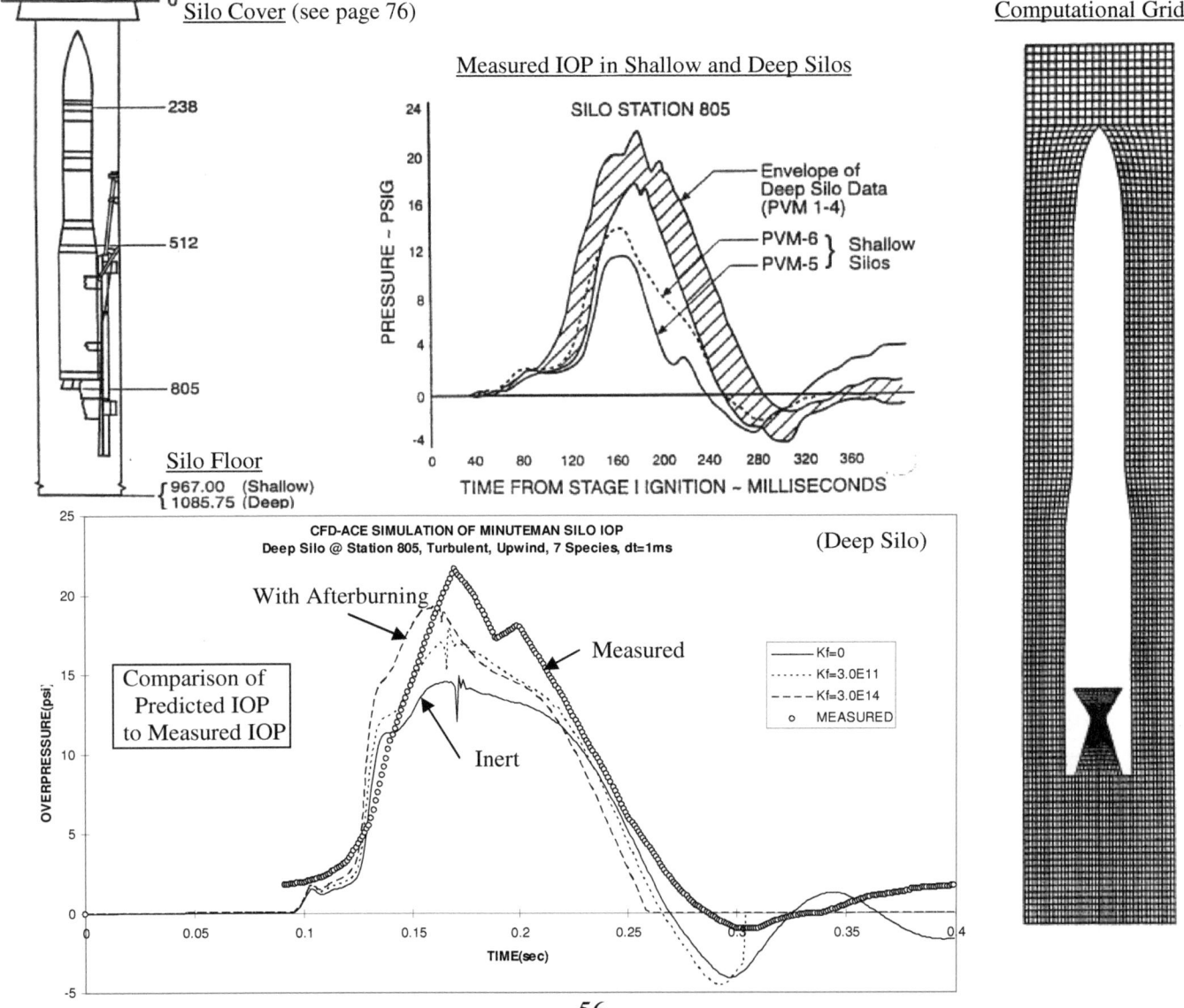

1D Propellant Ignition Code SHARPIT

The Problem: When the igniter fires, pressure waves are propagated down the combustion chamber (see page 25), and the spatial variation of chamber temperature has a large effect on propellant ignition. We needed to have at least a one-dimensional model of this process. The first 1D CFD model of propellant ignition was developed by Len Caveny and Ken Kuo at Princeton, but they didn't include an ignition model as sophisticated as mine, nor had it allowed for igniter injection forward of the head end of the chamber. We at Thiokol needed to acquire a 1D modeling capability; Jim Rozanski was to accomplish this very well.

The Solution: I had trained Thiokol engineer Jim Rozanski to use VOLFIL to model propellant ignition. As part of a Thiokol contract with AFRPL, Rozanski specialized the Thiokol axisymmetric CFD code SHARP to 1D, and added my VOLFIL ignition model. Some years later, I extracted his code from my copy of the AFRPL package and called it SHARPIT (Ref 49). It was to become very useful in modeling multiple ignition phenomena, as described below and later (pages 65 and 77).

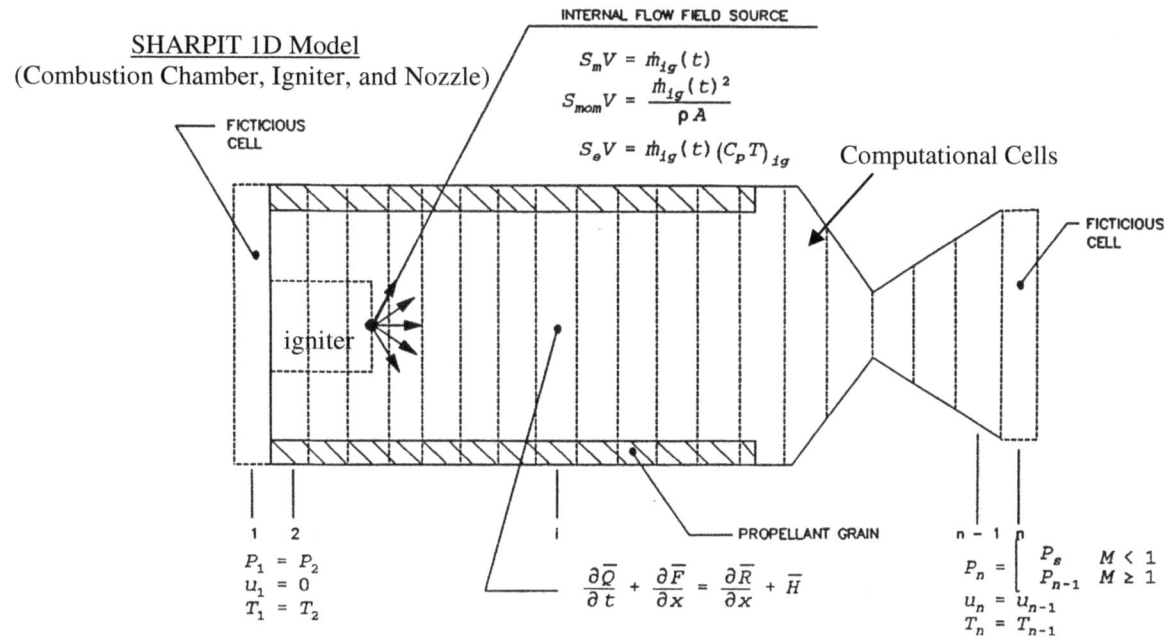

Neat Trick: During my modeling of Ignition Overpressure (IOP) during ignition of a Minuteman missile in a silo, I had an inspiration. I realized that chemically-inert SHARPIT should be able also to model the IOP in a minimally-afterburning shallow silo by interpreting (1) the silo as a combustion chamber with inert propellant (the silo walls), (2) the motor headend as the silo floor, (3) the nozzle throat as the open ground plane, and (4) by reversing the direction of the igniter exhaust to point at the headend of the motor (silo floor) to simulate the rocket exhaust.

> The resulting histories of IOP predicted by SHARPIT in a shallow silo are shown below to match surprisingly well the measured IOPs at both high and low silo stations. Estimates of peak IOP can now be made in seconds on a PC.

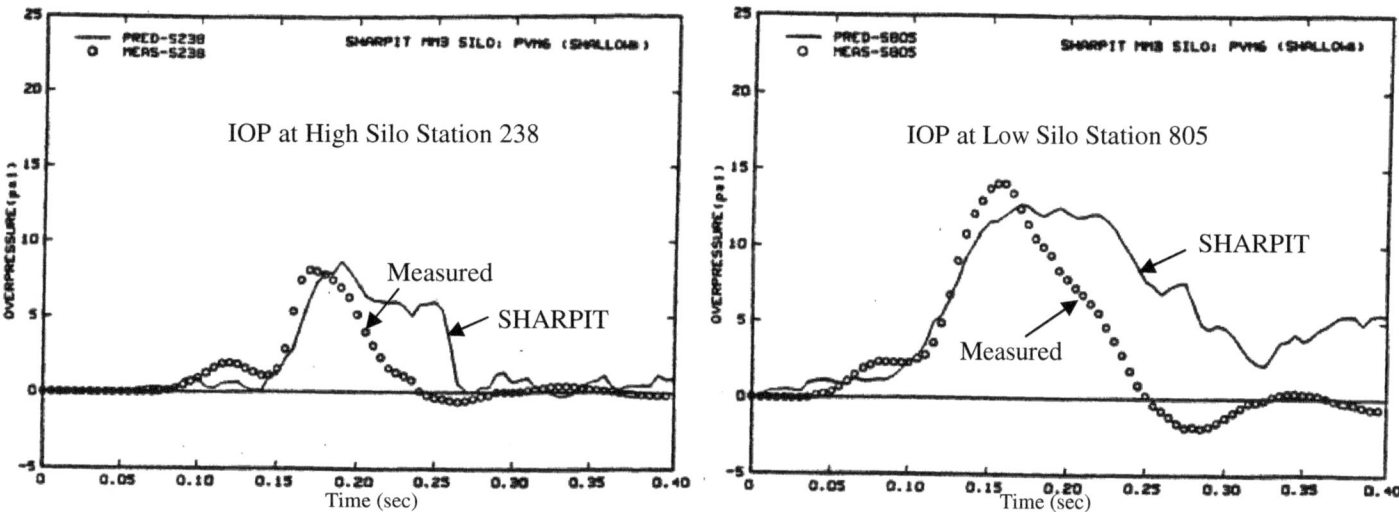

Due to my experience with propellant ignition, I had chaired a JANNAF workshop in 1985 (Ref 50), and subsequently an AIAA conference session in 2002 (Ref 51) dedicated to the subject.

CFD Model SALE3D of Vehicle Staging (Essentially An Ignition Event)

As discussed previously on page 24, vehicle staging begins when the upper stage motor ignites, and progresses as it and the lower stage separate a significant distance within about 150ms. Initially, only volume-filling modeling was available.

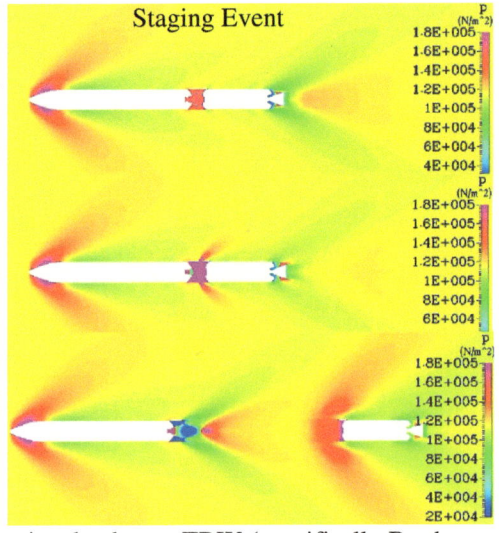

Staging Event

The Problem: Early in my tenure at Thiokol, the Air Force had issued an RFP (Request for Proposal) for the development of a CFD code to model the flow-field during vehicle staging. I called colleague Prof. Joe Hoffman at Purdue to see if he wanted to team with Thiokol in responding to this RFP. He said he'd prefer to bid alone since he could bid it much more cheaply due to the availability of cheap grad-student labor. Purdue did win the contract, and chose to modify the Los Alamos code SALE to allow for the time-dependent boundary conditions created by the stage separation dynamics. However, they were having difficulty getting the resulting code SALE3D to run properly. Indeed, at a Monterey conference, we sat in an ice cream shop while he bemoaned the fact that after many successful years of CFD modeling, he couldn't get the boundary conditions to work for the moving (expanding) grid in SALE3D. After that meeting, I lost track of the Purdue effort.

Lo and behold, years later when I joined TRW I found out that the Air Force had previously chosen TRW (specifically Brad Hazzard) to monitor and utilize the resulting successful Purdue code SALE3D. Brad was never able to get the 3D option to work, but the axisymmetric option ran fine. When Brad later announced that he was leaving TRW to work for Raytheon, I told my supervisor that someone in the group should learn how to use SALE3D, and that I was probably the best choice. He agreed, and I subsequently wrote a batch file to help operate the code and post-process the fairly complicated output.

My Solution: To help understand the staging event, I always present the following discussion (see Ref 1 or 24 for details):
- During pressurization of the upper-stage motor after ignition, a shock wave forms at the nozzle throat once it chokes, and propagates down the nozzle as chamber-to-cavity pressure ratio p_{cham}/p_{cav} increases
- The shock reverses direction back toward the throat as cavity pressure temporarily grows faster than chamber pressure, then moves aft again as the stages separate and the cavity vents
- During this process the shock temporarily pops into a skewed (asymmetric) orientation due to circumferential variation in the location of separation of the nozzle boundary layer, yielding large nozzle side force
- This side force could damage actuators or push the nozzle sideways into the upper-stage skirt (bad), so predicting peak nozzle "sideload" is important
- Eventually the shock pops out into the cavity, but may temporarily be swallowed back into the nozzle if venting of the interstage cavity occurs slowly

Three phases can be shown schematically:

Propagation of Upper-Stage Start-Up Shock Down Nozzle and Into Interstage Cavity, with Temporary Asymmetry

| Choked Throat/Unvented Cavity (Symmetric Shock Propagates Down Nozzle) | Cavity Venting Begins and Temporary Shock Skewing (Nozzle Side Loads) | Full Cavity Venting Shock Pops into Interstage Due to Low Cavity Pressure |

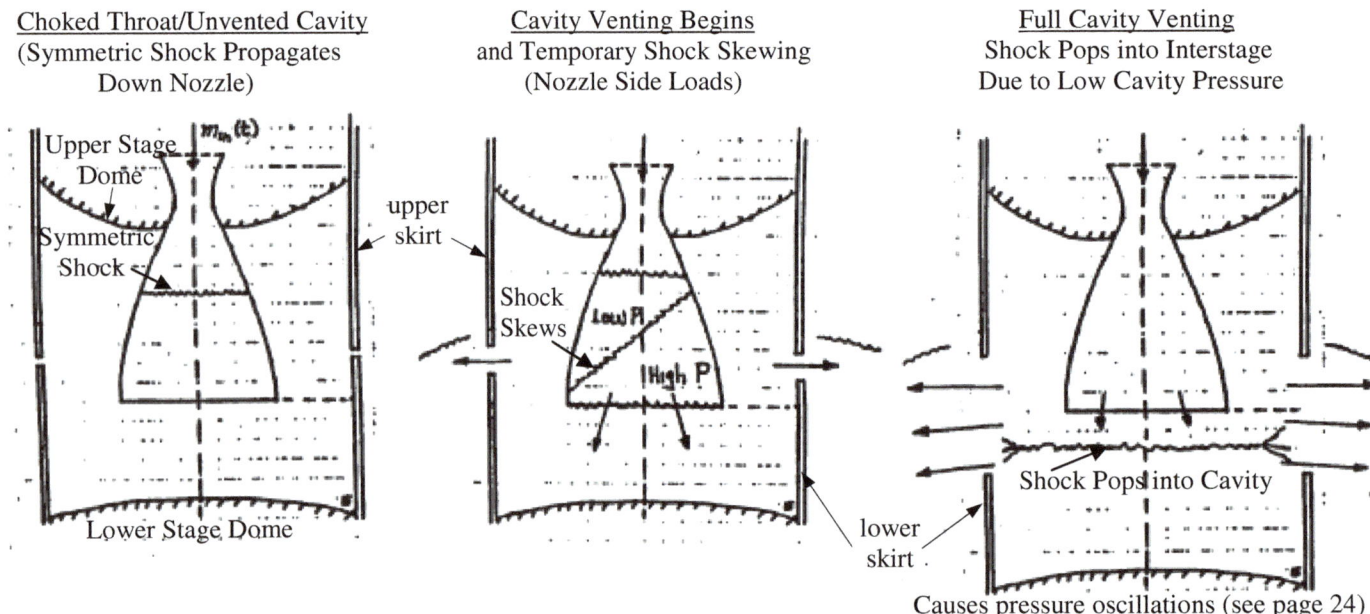

Causes pressure oscillations (see page 24) due to fluctuation of shock near lower skirt.

I modeled the staging event for a number of vehicles using both SALE3D and my volume-filling code STAGING (see page 24) with both vented and unvented skirts. Some results using SALE3D for Minuteman 1/2 staging are shown below:

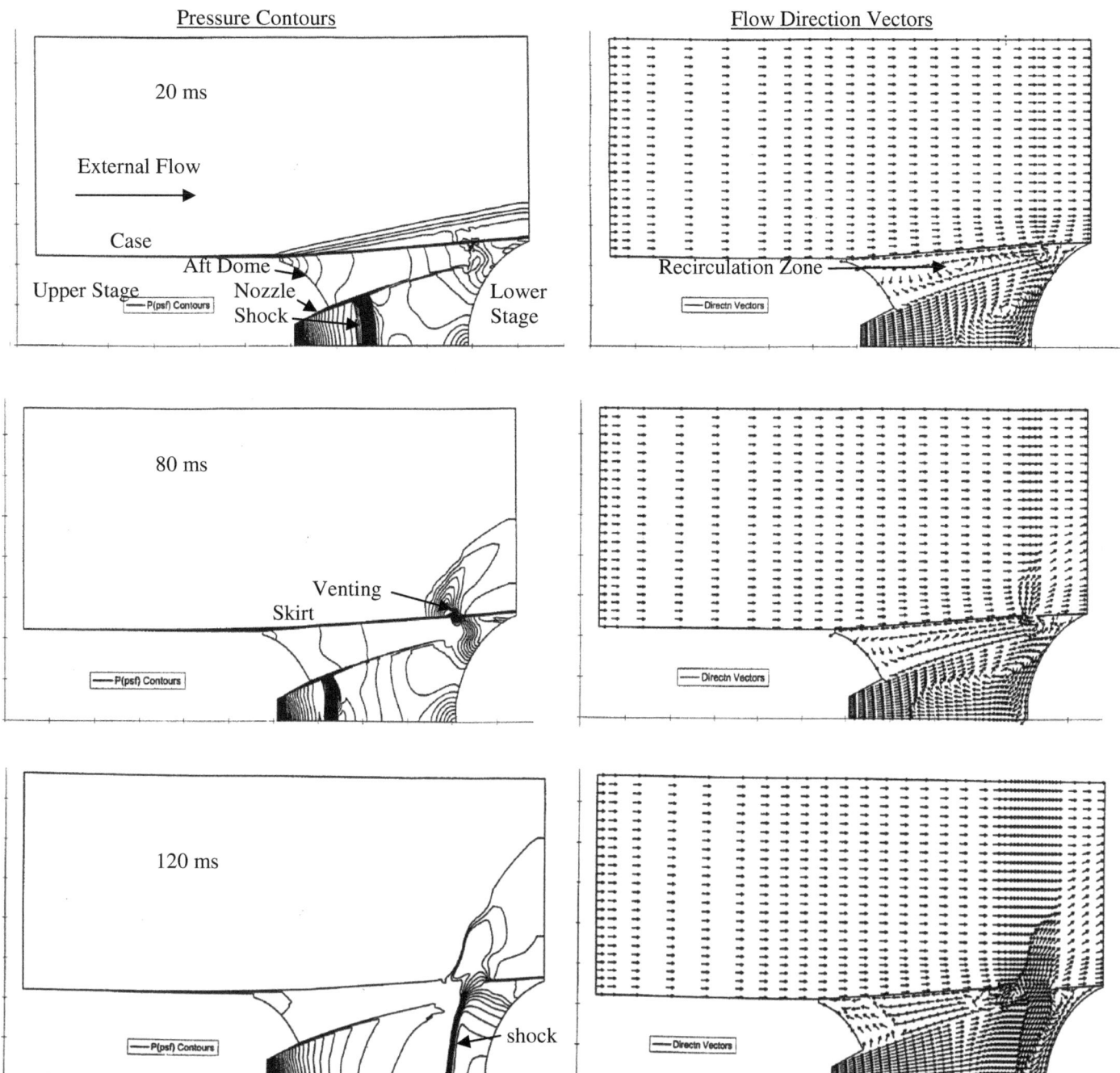

Pressure Contours   Flow Direction Vectors

Nozzle Sideload

The peak side load on the nozzle during staging is of major concern. It has been known to rupture the actuators that rotate (gimbal) the nozzle for thrust vector control (TVC). I subsequently developed a prediction method (1) using SALE3D to predict the symmetric propagation of the startup shock down the nozzle and into the interstage cavity, and then (2) creating a semi-empirical code SIDELOAD in an attempt to estimate shock skewing and the resulting peak nozzle side force during motor ignition. It generated agreement within 30% for Minuteman staging events. However, the peak side load has been known to vary by a factor of five among staging events of supposedly identical vehicles (Peacekeeper, SRMU). This is caused by non-repeatable boundary-layer separation in the upper stage nozzle. Consequently, success in modeling peak side loads has proven elusive, even when using the best available CFD codes.

I was subsequently asked by JANNAF to chair a workshop on staging and sideloads in 2002 (Ref 54).

(5E) Making Codes User-Friendly

The Problem: The development of many computer programs critical to the U.S. rocket industry was funded by the federal government (Department of Defense). Unfortunately, most of that funding went to develop the technology, and little went to make the codes user-friendly. I call the resulting programs "the codes from hell" due to their horrible or non-existent user interfaces. However, the codes are valuable technically, so I have made a special effort over the last several decades to make numerous government-sponsored codes user friendly. Some such PC codes are:

CMA92, 04 and ACE	Aerotherm Charring and Material Ablation versions 1992 and 2004
FIAT	NASA Ames Charring and Ablation Code
MINIVER, EXITS	External Aerodynamic Material Heating and Ablation
SPF-III v 4.0 and 5.0	Standard Plume Flowfield Code
SPURC v1.3	Standard Plume Ultraviolet Radiation Code
NARJ, PRCJ	Russian Numerical Analysis of Real Jets (Nozzle/Plume Flowfields and Signature)
OD3P	One Dimensional Three-Phase Nozzle Flow Model
MN2IT	Blunt Nosetip Flowfield Generator for Starting Supersonic Aero Codes
ZEUS	Supersonic Aerodynamic Prediction Code
NPARC	National Propulsion Aero Research Code

My Solution: The main factors that allowed me to convert a "code from hell" into a user-friendly code package were:

1) I learned how to write a Menu-Driven DOS Batch File to control the operation of the code and its pre- and post-processing (Input/Output), as demonstrated on the following pages.
2) I developed my EXCEL and MATLAB plot packages, which provide most of the required plotting capability for any code (mostly scatter plots, multizone grids, contour lines, velocity and direction vectors, color-bands).

The basic steps I take when I acquire a new user-unfriendly computer code (even 7 years into "retirement") are:

1) If the source code is provided, (a) the input file is converted to Namelist format if it is not already, and (b) the format of the output plot files must be modified to be consistent with my plot package requirements, including title lines and column headers with units (too many original plot files even had no headers ... a very bad practice).
2) If the source code is not provided, an interface processor must be written to read the original run output file, and extract and convert the appropriate output to the required plot format.
3) Clone an existing batch file, and modify it for use with the new code. This generally takes between 5-30 minutes, and often requires only (a) changing the code name in the batch file, (b) changing the paths to the directory where the code will reside, and (c) changing the names of the input and output files to be processed by the batch file.
4) The bulky detailed input and output description provided in the supplied User's Guide is converted into a stream-lined on-line User's Guide, typically one line per variable. All the input parameters and their required units and inter-relationships are now easy to identify; this and all supplied User Manuals can be opened from the batch file.

My development of user-friendly versions of three codes for liquid and solid rockets is described on the next few pages.

Menu-Driven DOS Batch Files are Better Than GUIs at Automating Computer Code Operation

Unfortunately, the menu-driven DOS batch file has been overlooked by scientists and engineers as a tremendous way to make the operation of one or a set of PC computer programs **User-Friendly**. It is equivalent to a GUI without the graphics (the menu choices are equivalent to clicking a GUI icon or dialog box), but has the following advantages:

1) It is Easier to Create and Clone a Batch File: Examples are shown on pages 51, 61, 63, 64, and 90-91.
   It is Easier to Insert On-Line Instructions at Each Step: Allows user customization any time with ECHO command.

2) It is Easier to Modify a Batch File to Cycle the Execution Automatically.
   Suppose your code runs for only one set of parameters. A short post-processor can be written to copy the baseline input file to a new file and modify any necessary parameters to allow a new run to be made. Several examples created by this author:
   a) A solar heat flux attenuation code MODTRAN written for only one time-of-day (single solar aximuth) was cycled to run automatically in one-hour (or any desired) increments to generate the entire diurnal cycle.
   b) The 1D Charring Material and Ablation Code (CMA) was cycled automatically to 50 axial locations along a nozzle wall to generate a quasi-two-dimensional thermal solution along the nozzle (Ref 55).

3) It is Easier to Create a Batch File to Run a Sequence of Codes or a Select Set of Codes (see EVT example, page 51).

4) It is Easier to Create a Batch File to Run a Code Parametrically for a Series of Flowfields.
   Calculations of station radiant intensity in plumes were made with SPURC for more than 400 gas/particle radially-uniform flowfields (Ref 73). Each case had to be run individually, but the batch file RUNSPURC automated the process using Menu Option #8 so well that each case could be set up and run in less than two minutes.

Example 1: Making the Charring and Material Ablation Code CMA User-Friendly

The Problem: CMA is very useful for predicting the 1D heating and ablation of materials at a single location, but the input and output formats of the original versions were horrendous. When AFRL thought that they would acquire some extra funding, they asked industry to suggest some projects that they could support. I suggested that I could make CMA user-friendly, and they thought that was a great idea. Unfortunately, AFRL never received the expected funds, so they said "sorry" to me.

My Solution: However, TRW and I wanted a usable version of CMA for ourselves, so I decided to make the effort anyway. It took me three tries over a period of five years to convert the 1992 version to a user-friendly version CMA92 (Ref 56). It was modified to read input in a user-friendly format: (1) input is now read in NAMELIST form rather than the original fixed-column format, (2) material thermal properties (including decomposition and pyrolysis data) are read from an industry-standard data file MATPROP that enhances user-to-user consistency and greatly reduces user input effort and potential error, (3) nodal data is specified much more compactly than the original form, and (4) a menu-driven DOS batch file RUNCMA was created to automate the operation of the code (see partial listing below).

In addition, plot files were written that allow automatic plotting of thermal and recession output when CMA92 is run from RUNCMA (see page 47). A number of problems were simulated to verify that both formats generated identical solutions, from a single non-decomposing main material, to one with two decomposing and two non-decomposing backup materials.

The Namelist input format also allowed me to create an interface CMACYCLE that could run CMA92 sequentially and automatically for up to 50 points along a nozzle wall in a single run (Ref 55). The main menu in RUNCMA is listed below:

```
MAKE CHOICE FOR "%INFILE%"
        --- CMA92 ---                              OUTPUT FILES AVAILABLE FOR "%INFILE%" ...
    0) Quit
    1) Edit input  file "%INFILE%.INP"             0) Return to menu
    2) Run  CMA92                                  1) Run-summary file "%INFILE%.OUT"
    3) Process output files
                                                   Edit/Plot:
        --- CMACYCLE ---                           2) File 2   (energy-balance histories)
    4) Generate cycle file from                    3) File 3   (recession data)
            (a) nozzle    definition file FNOZ     4) File 4   (temperature histories at surface and thermocouples)
        and (b) heat flux definition file FBND     5) File 16  (temperature profiles at up to 15 specified times)
    5) Cycle CMA (iff CYCLEFIL available from option 4)
                                                   Edit only:
        --- UTILITIES ---                          6) File 7   (temperature histories at NI isotherm depths)
    6) Run ACE (not yet hooked in)                 7) File 11  (in-depth temperature profiles in sequence)
    7) Browse surface thermochemistry table        8) Surface Thermochem Data file  "%INFILE%.STC"
    8) Browse MATPROP.DAT
    9) Browse User's Guides

        --- CYCLE POST-PROCESSING ---
    A) Generate merged file by rays of F21 thermal profiles          (after option 5)
    B) Generate merged file by time of F21 thermal profiles for ABAQUS (after option A)
    C) Generate merged file by time of F03 recession + char histories (after option 5)
    D) Generate merged file by time of F04 wall temperature histories  (after option 5)
    L) Generate merged file by time of T at last nodes from F21 file   (after option A)
    E) View individual file \CMA92\OUTPUT\%INFILE%_type.N
    F) List B'c, B'g, Twall at all times from specified CMA output file
    G) Generate material color plot from CMACYCLE.PLT
```

The new input format reduced the original 121-line fixed-format input file to the 31-line Namelist file listed below:

```
TEST CASE #2: CARBON-PHENOLIC MAIN + SILICA-PHENOLIC DBU + TITAN NDBU
Original format replaced by namelists and data bases:
  Namelists : FLAGS, MATER, MARCH, NODES, DECOM, BOUND
  MATPROP.DAT: Main material (815), DCBU mat'l (31), NDBU mat'l (207)
  THERMOFILE : FCHEM=ARIES.TAB
?
*MISCELLANEOUS FLAGS AND CONSTANTS
 &FLAGS  ICNTR=0, IVAC =0, NCON =1,     NFIS=0, IPGPRS=0,
         NN   =1, IHTCF=0, NOPT1=1,     NR =0, NST   =0,          &END
 &MATER  NDBU =1, NFFUNC=0, DELM =0.5,  RA = 0.0, 0.0, JF=2,
         HCONV=0.0, EPSW=0., TRES=0.,
         CMHS =0.6753, VFZ=0.888, BRP=0.4,
         CHCRI=0.27, PYCRI=0.0, BREX=0.0, SWELL=0.0,              &END
 &MARCH  THZRO=10.0, THFIN=56.0, DTHB=1.0,
         NO=4, NI=3, SO=0.10, 0.25, 0.35, 0.45, 2000., 1460., 1000.,
         NOUT =3,   ADTPRT=3*0.1, ATPR=1.0, 55.0, 70.0,
         TPLOT=20.0, 30.0, 40.0, 50.0,                            &END
*DECOMPOSITION KINETIC DATA
 &DECOM MATTAB='MATPROP.DAT', MAIN=815, MATDBU=31, MATNON=207,    &END
*NODAL DATA
 &NODES MATL=13*1, 5*22, 3*3, TA=21*520., AREA=21*0.,
        DEL=0.010, 18*0.020, 0.030, 0.050,                       &END
*SURFACE BOUNDARY CONDITIONS
 &BOUND TTH=0.0, 5.06, 10.05, 15.04, 20.03, 25.03, 30.0, 35.0,
           40.0, 45.0, 50.05, 55.04, 60.03, 65.0, 70.0,
        THE=15*1000.,
        TQR=3*3.7, 3.8, 3.9, 4.1, 4.4, 4.8, 5.3, 5.8, 6.9,
           7.9, 7.9, 7.6, 2.7,
        TCM=15*0.0,   TPI=15*32.9,   TBRP=15*0.0,   TABLE=0,      &END
*SURFACE THERMOCHEMISTRY TABLES
 &SURFT FCHEM='\CMA92\ARIES.TAB', FORMAT=1,                       &END
----------------------------------------------------------------------
```

Example 2: Generating a User-Friendly Version of the Standard Plume Flowfield Code SPF

The Problem: Modeling the flowfield of a rocket exhaust plume is important for predicting the heating of nearby structures and the aft end of the vehicle, and as input to codes that predict plume signature for target discrimination. The JANNAF Standardized Plume Flowfield Code SPF-III Version 4.0 was obtained from CPIA and converted from UNIX/SGI form to run on a PC. Although Version 4.0 was reported to be much more user-friendly than Version 3.5, it was still very unfriendly. This unfriendliness was due to the complex code structure (4 separate executables for four separate flow modules), the 7-domain shock fitting methodology, poor coding practices of the many organizations that had been involved in developing the code over nearly two decades, and the poor management of the nearly 30 input/output files.

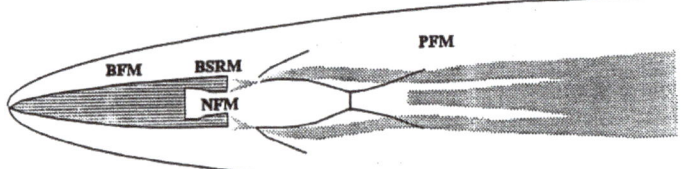

My Solution: I made many modifications (Ref 57) that made SPF-III much more user-friendly. The biggest improvements were (1) to unify the nozzle start-plane input from the original 3-file system (separate files for gas, chemical species, and droplet properties) to a single unified file containing all three sets of properties, (2) to simplify the operation of the code package by creating the menu-driven DOS batch file RUNSPF (shown on the next page), and (3) to modify the format of the output plot files to match the requirements of my EXCEL plot package, which is run automatically by the batch file.

Typical results are shown below for a plume emanating from a diverging nozzle: (a) a Schlieren photograph of the shock train downstream of the nozzle exit, (b) the temperature behavior along the plume axis, and (c) the temperature color-bands:

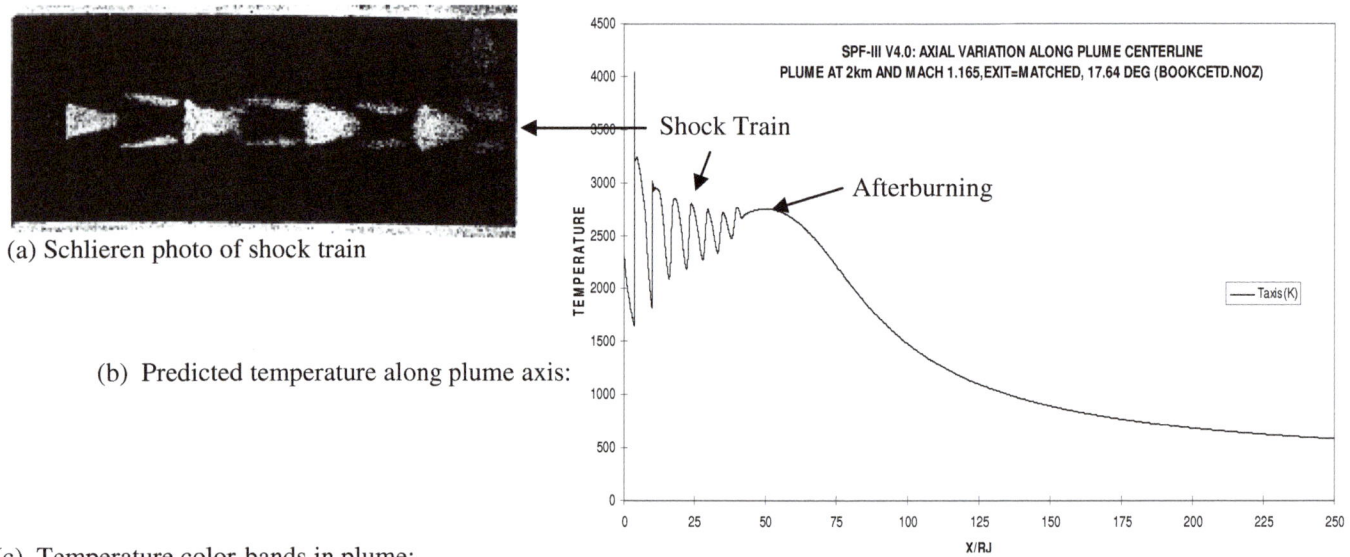

(a) Schlieren photo of shock train

(b) Predicted temperature along plume axis:

(c) Temperature color-bands in plume:

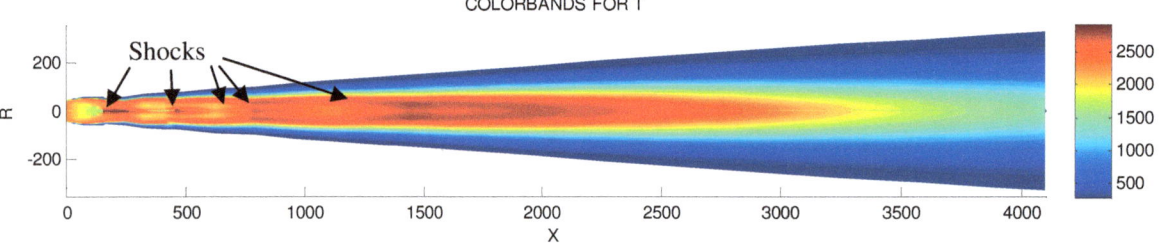

The many input and output files just for the Plume Flowfield Module (PFM) of SPF-III are now managed by RUNSPF:

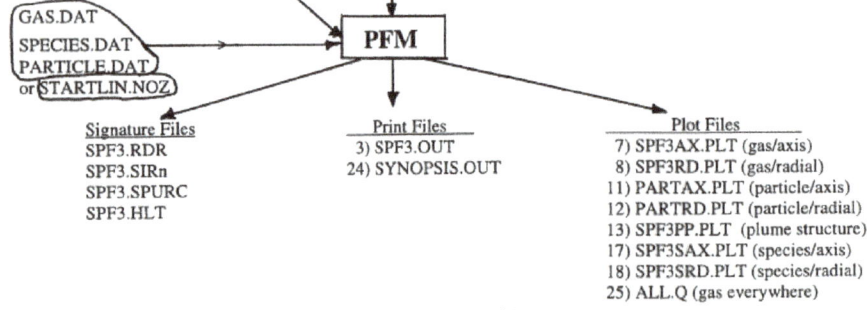

Batch File Menus for Plume Code SPF-III V4.0:

The flow chart below may look complicated, but it really is simple to navigate since only one menu at a time appears on the PC monitor. Note how <u>all options available to the user are listed</u>, and each step in the process follows in sequence. Note also how easy it is to choose the appropriate plot, and to be reminded of the file name. Thus, the user is led by the hand using RUNSPF, rather than left in the cold using the original version of the code package.

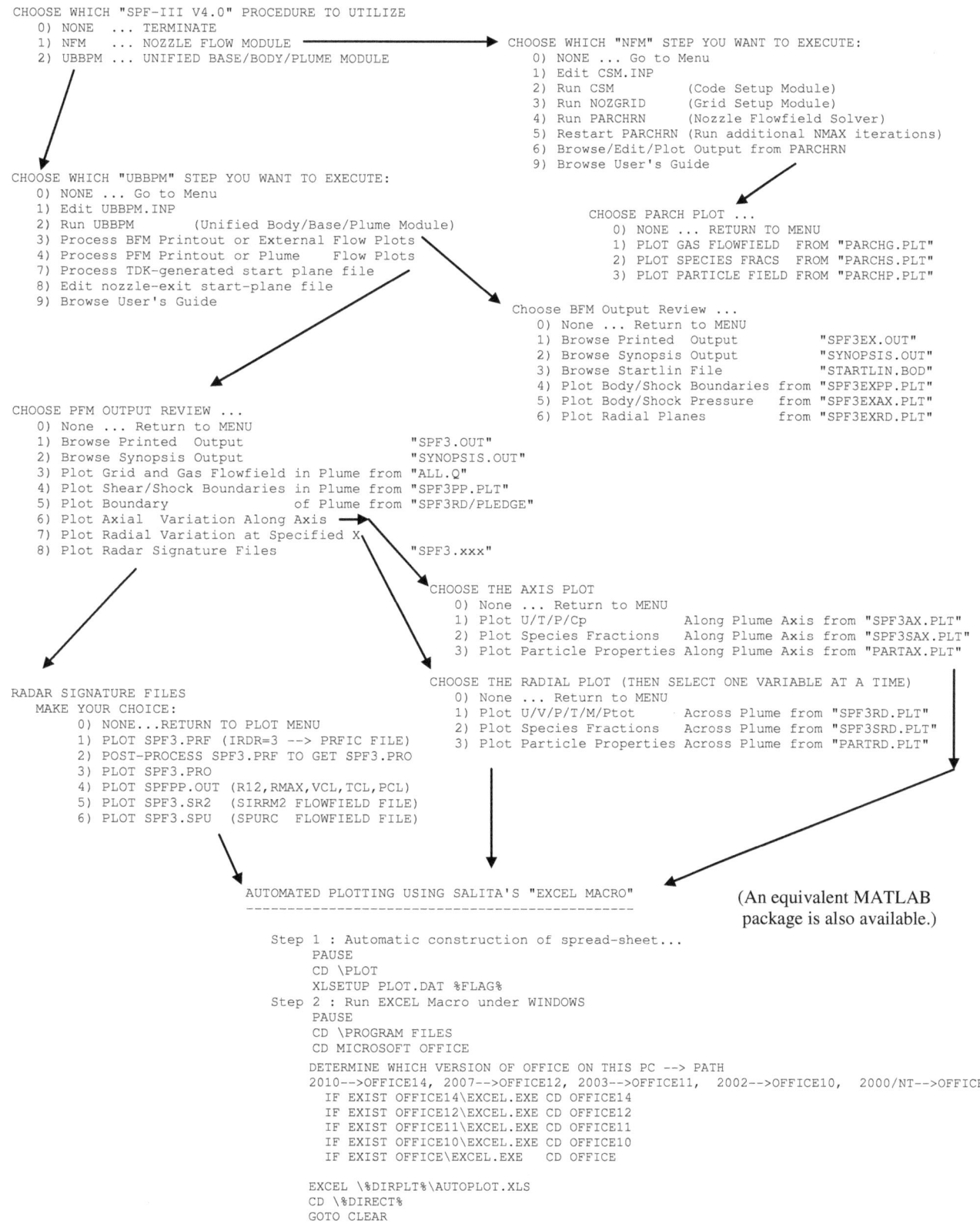

```
CHOOSE WHICH "SPF-III V4.0" PROCEDURE TO UTILIZE
   0) NONE  ... TERMINATE
   1) NFM   ... NOZZLE FLOW MODULE
   2) UBBPM ... UNIFIED BASE/BODY/PLUME MODULE
```

```
CHOOSE WHICH "NFM" STEP YOU WANT TO EXECUTE:
   0) NONE ... Go to Menu
   1) Edit CSM.INP
   2) Run CSM          (Code Setup Module)
   3) Run NOZGRID      (Grid Setup Module)
   4) Run PARCHRN      (Nozzle Flowfield Solver)
   5) Restart PARCHRN (Run additional NMAX iterations)
   6) Browse/Edit/Plot Output from PARCHRN
   9) Browse User's Guide
```

```
CHOOSE WHICH "UBBPM" STEP YOU WANT TO EXECUTE:
   0) NONE ... Go to Menu
   1) Edit UBBPM.INP
   2) Run UBBPM       (Unified Body/Base/Plume Module)
   3) Process BFM Printout or External Flow Plots
   4) Process PFM Printout or Plume   Flow Plots
   7) Process TDK-generated start plane file
   8) Edit nozzle-exit start-plane file
   9) Browse User's Guide
```

```
CHOOSE PARCH PLOT ...
   0) NONE ... RETURN TO MENU
   1) PLOT GAS FLOWFIELD  FROM "PARCHG.PLT"
   2) PLOT SPECIES FRACS  FROM "PARCHS.PLT"
   3) PLOT PARTICLE FIELD FROM "PARCHP.PLT"
```

```
Choose BFM Output Review ...
   0) None ... Return to MENU
   1) Browse Printed  Output          "SPF3EX.OUT"
   2) Browse Synopsis Output          "SYNOPSIS.OUT"
   3) Browse Startlin File            "STARTLIN.BOD"
   4) Plot Body/Shock Boundaries from "SPF3EXPP.PLT"
   5) Plot Body/Shock Pressure   from "SPF3EXAX.PLT"
   6) Plot Radial Planes         from "SPF3EXRD.PLT"
```

```
CHOOSE PFM OUTPUT REVIEW ...
   0) None ... Return to MENU
   1) Browse Printed  Output                   "SPF3.OUT"
   2) Browse Synopsis Output                   "SYNOPSIS.OUT"
   3) Plot Grid and Gas Flowfield in Plume from "ALL.Q"
   4) Plot Shear/Shock Boundaries in Plume from "SPF3PP.PLT"
   5) Plot Boundary           of Plume from "SPF3RD/PLEDGE"
   6) Plot Axial  Variation Along Axis
   7) Plot Radial Variation at Specified X
   8) Plot Radar Signature Files             "SPF3.xxx"
```

```
CHOOSE THE AXIS PLOT
   0) None ... Return to MENU
   1) Plot U/T/P/Cp          Along Plume Axis from "SPF3AX.PLT"
   2) Plot Species Fractions  Along Plume Axis from "SPF3SAX.PLT"
   3) Plot Particle Properties Along Plume Axis from "PARTAX.PLT"
```

```
RADAR SIGNATURE FILES
   MAKE YOUR CHOICE:
      0) NONE...RETURN TO PLOT MENU
      1) PLOT SPF3.PRF (IRDR=3 --> PRFIC FILE)
      2) POST-PROCESS SPF3.PRF TO GET SPF3.PRO
      3) PLOT SPF3.PRO
      4) PLOT SPFPP.OUT (R12,RMAX,VCL,TCL,PCL)
      5) PLOT SPF3.SR2  (SIRRM2 FLOWFIELD FILE)
      6) PLOT SPF3.SPU  (SPURC  FLOWFIELD FILE)
```

```
CHOOSE THE RADIAL PLOT (THEN SELECT ONE VARIABLE AT A TIME)
   0) None ... Return to MENU
   1) Plot U/V/P/T/M/Ptot    Across Plume from "SPF3RD.PLT"
   2) Plot Species Fractions  Across Plume from "SPF3SRD.PLT"
   3) Plot Particle Properties Across Plume from "PARTRD.PLT"
```

(An equivalent MATLAB package is also available.)

```
AUTOMATED PLOTTING USING SALITA'S "EXCEL MACRO"
-----------------------------------------------
Step 1 : Automatic construction of spread-sheet...
         PAUSE
         CD \PLOT
         XLSETUP PLOT.DAT %FLAG%
Step 2 : Run EXCEL Macro under WINDOWS
         PAUSE
         CD \PROGRAM FILES
         CD MICROSOFT OFFICE
         DETERMINE WHICH VERSION OF OFFICE ON THIS PC --> PATH
         2010-->OFFICE14, 2007-->OFFICE12, 2003-->OFFICE11,  2002-->OFFICE10,  2000/NT-->OFFICE
            IF EXIST OFFICE14\EXCEL.EXE CD OFFICE14
            IF EXIST OFFICE12\EXCEL.EXE CD OFFICE12
            IF EXIST OFFICE11\EXCEL.EXE CD OFFICE11
            IF EXIST OFFICE10\EXCEL.EXE CD OFFICE10
            IF EXIST OFFICE\EXCEL.EXE   CD OFFICE

         EXCEL \%DIRPLT%\AUTOPLOT.XLS
         CD \%DIRECT%
         GOTO CLEAR
```

Note how easily the user can tailor the on-line instructions for each step in the process…better than a GUI !!!

Example 3: Making the Aero Code Prediction Package ZEUS User-Friendly

The Problem: My supervisor gave me a copy of the 3D supersonic inviscid flowfield prediction package ZEUS (1986 version) for tactical missiles with fins, and asked me if I could make the code package user friendly. ZEUS is a collection of three codes to calculate the forces on a tactical missile with fins in supersonic flight at angle of attack (more on page 68):

(1) a nosetip flowfield simulator (for pointed noses this is INIT) to generate a start-plane for downstream calculations (I subsequently fixed a major error in INIT for angle of attack),

(2) an inviscid solver (ZEUS) which marches the flowfield calculation axially from the start-plane to the tail, and

(3) a code (CONVERT) to rezone the computational grid for axial domains containing fins.

Unfortunately, the versions of these codes previously in use were very user unfriendly because

(1) body and fin geometries had to be coded by the user into the source coding for both ZEUS and CONVERT each time a different missile was modeled (incredibly awkward),

(2) different sets of these geometry routines had to be linked to ZEUS and CONVERT,

(3) ZEUS and CONVERT required different input files even though most of the input data was the same,

(4) the codes had to be manually executed in sequence with restart files (INIT, then ZEUS, then CONVERT and ZEUS),

(5) the theory and user manuals were poorly written and hard to understand, and

(6) no validation cases were provided (predicted versus measured force coefficients).

My Solution: I generated a modified version ZEUS97 (Ref 58) and automated the INIT/ZEUS97/CONVERT/ZEUS97 simulation process:

(1) A single include file now provides array dimensions for both ZEUS97 and CONVERT.

(2) A generalized definition of body and fin geometry (including thickness) was built into the single subroutine bundle ZEUSG which provides all necessary geometric models for both ZEUS97 and CONVERT.

(3) The required input geometric parameters were specified by the user in a single input file for all ZEUS97 and CONVERT runs.

(4) A start-plane generator MN2IT (see page 69) was modified to replace INIT for blunt noses.

(5) Plot files were written in the forms expected by my plot package to verify correct geometry input and flow solution.

(6) The sequential execution of the codes was automated via menu-driven DOS batch file RUNZEUS:

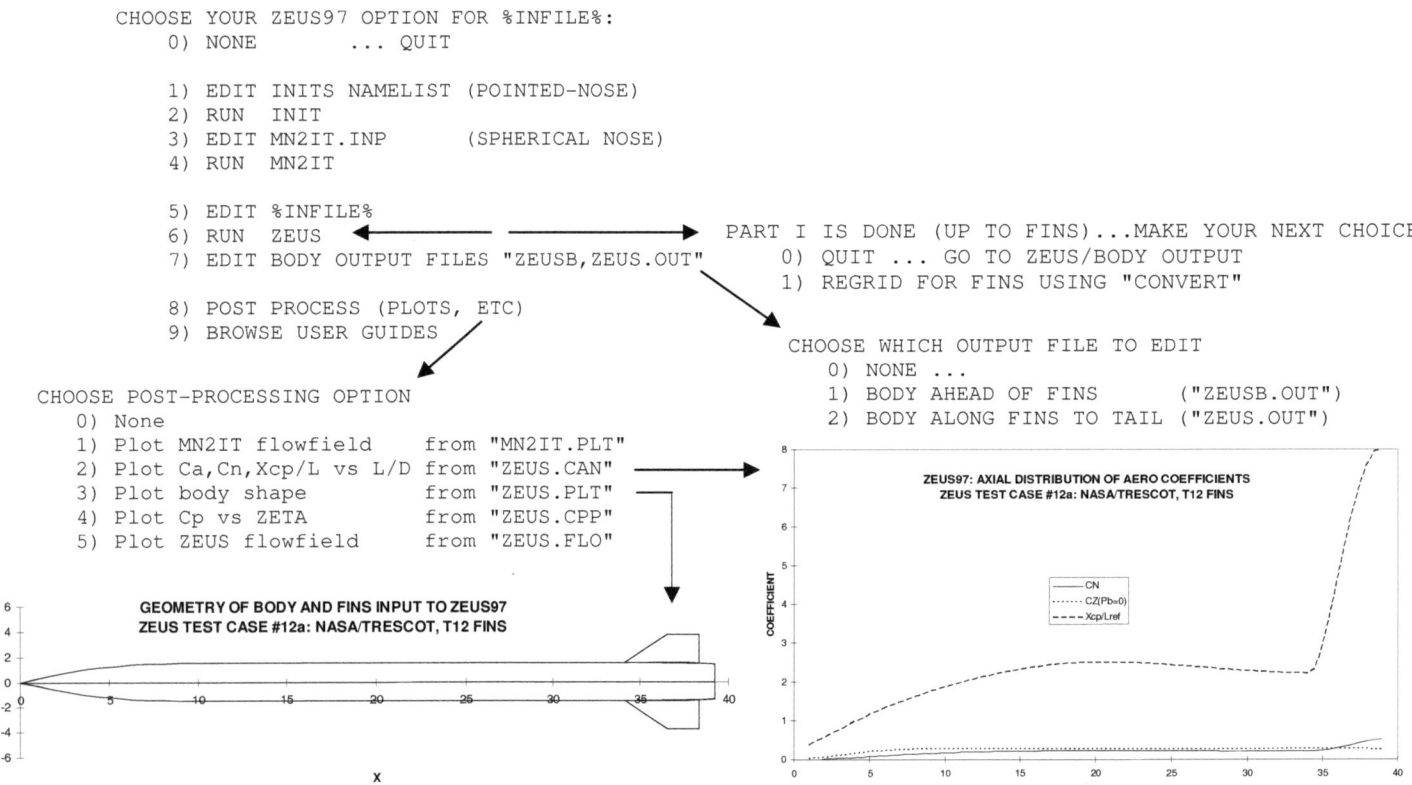

When I asked my former colleagues at Thiokol if they used ZEUS, they said that they had it, but it was a "bear" to run. I told them that I had generated a user-friendly version. They said "Please, please, can we get a copy?" I asked my TRW management if I could send a copy to Thiokol. They replied in the negative, since TRW had spent the time/money for me to make it user friendly.

(5F) Lectures Here and Abroad: Cashing In On What I'd Learned

Paris Lectures to ONERA

In static tests, the French solid-propellant rocket booster Ariane 5 was found to have a huge amount of slag (≈2000 kg), comparable to that in Titan SRMU tests. After reading my slag article in the February 1995 issue of Journal of Propulsion and Power (Ref 4), ONERA (France's equivalent of NASA) asked NATO's scientific arm AGARD to invite me to lecture to them about slag generation. NATO would pay my expenses plus an honorarium.

I agreed and within two months, my wife and I flew to Paris (although our intended arrival on Saturday was foiled by aircraft mechanical problems in Salt Lake City). We arrived Monday morning, and my liaison Guy Langelle picked us up at the airport, drove us to our hotel in the Latin Quarter, waited while I changed my clothes, then drove him and me to ONERA, where I fought jet lag to present the first of my 3-day lectures. The rest of the week my wife and I toured Paris and Versailles.

Brussels Lectures at the Von Karmen Institute

I was invited by Vigor Yang (then a professor at Penn State) to participate in a lecture series to be given at the Von Karman Institute in Brussels in May 2002. The lectures were titled "Internal Aerodynamics in Solid Rocket Propulsion". I basically presented my AIAA paper 'Predicted Slag Deposition Histories in Eight Solid Rocket Motors Using the CFD Model "EVT"', as well as a summary of the Thiokol Quench Bomb data. Other U.S. lecturers were Fred Culick of Caltech, John Murdock of Aerospace Corporation, and Merrill Beckstead of BYU. ONERA lecturers included Langelle, Vuillot, Guery, and Kuentzmann. In the evenings, some of us and several of the students (all graduate engineers, including several Brits) went together to fine restaurants and a jazz club. The conversation (luckily all in English) was very stimulating. One of the students had to fly the next evening to perform a flute concerto with the Toulouse Symphony, then return in time for the morning lectures. Ironically, my presentation on slag generation followed Beckstead's presentation on aluminum combustion, just as it did previously at an AIAA seminar several years earlier, and would seven years later in Israel.

CSAR Lecture at the University of Illinois

The Center for Simulation of Advanced Rockets (CSAR) won a contract with the Department of Energy (DOE) to demonstrate a massive parallel-computing capability. They accomplished that demonstration by applying it to CFD/Structural Analysis of the Shuttle booster operation. Incidently, their name was incorrect since the rocket was not advanced; it was the simulation that was advanced. Hence, the name should have been CASR (Center for the Advanced Simulation of Rockets).

The Problem: They published many papers over multiple years, some by graduate students. I criticized an early one on Ignition Transient Modeling because erroneous choices of parameter values led them to make erroneous conclusions; I notified them of their mistakes. A year or two later, another graduate student made similar mistakes in an AIAA paper.

My Solution: I contacted CSAR again, noting that they had repeated their errors. Meanwhile, their Oversight Committee, aware of these errors, recommended that I be brought in to present "A Dose of Reality". I subsequently spent two days in October 2000 lecturing to them about the Shuttle booster.

In particular, they couldn't predict correctly the "knee" in the pressure history during motor ignition, i.e., where the pressure rise rate halves. This was not uncommon; an earlier Aerospace Corporation paper had shown that their predicted transient disagreed with the measured booster data, and pronounced that "**the data must be wrong**". It wasn't!!!

Well, the data was correct and consistent over many firings, and was matched well by the Thiokol **SHARPIT** code (page 57), which showed the reflection of the igniter shock from the aft end of the booster (see the hoop strain gauge data on page 25); the knee occurs when this shock reflection from the aft end then reflects from the head-end pressure transducer (see also the waterfall plot on page 48).

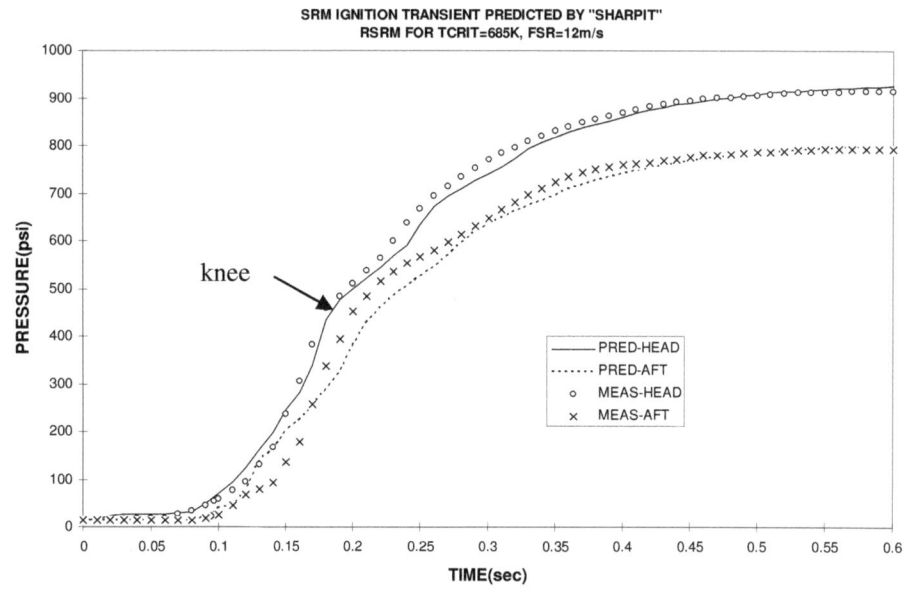

65

(5G) Predicting Aerodynamic Coefficients of Vehicles in Supersonic Continuum Flow

Prediction of the forces and moments on vehicles in flight is not an easy task. Flight velocities may be subsonic or supersonic, and the vehicles often have fins for control. Viscous effects and body heating can become important.

There are three basic types of computer codes to predict aero coefficients: (1) simplified engineering methods like Newtonian theory, (2) CFD methods are the most general but most difficult and expensive, and (3) component buildup methods that piece together analytical and empirical methods separately for bodies and fins and even exotic body shapes, but usually require a lot of artistry rather than formal modeling.

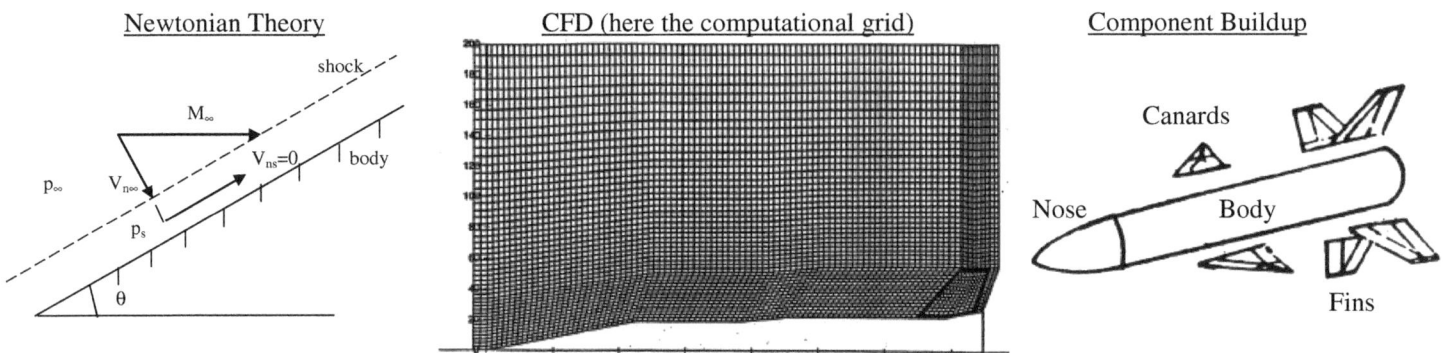

The Problem: Different aero codes often predicted different values of axial and normal forces, so we often ran multiple codes and compared the results. Other analysts in my group ran sophisticated CFD codes like CFD++, EULERNEQ, or GASP. For comparison, I was often asked to run ZEUS97 and a leased commercial code CFD-ACE, as well as component buildup codes (AP98 and DATCOM07) and my suite of simplified methods (NEWTON, SCONE).

> **My Solutions**: In my rocket career, I've always tried first to find simple but approximate solutions to problems. They expose functional dependence and often actually generate first-order-accurate quantitative solutions. Sometimes, when answers are needed quickly, that has to suffice. If more time is available and more accuracy is required, then it's time to go to the next level, and if necessary a third level. In the end, it often depends on the complexity of the vehicle shape, the available funding, and/or the required solution date.

Level 1: Approximate Engineering Solutions

Simplified engineering methods are primarily possible only for supersonic flight of simple geometric vehicle shapes, e.g., cylindrical bodies with conical, ogival, or blunt-cone nosetips, and planar fins. Tabulations of numerical solutions are available in classical publications like NACA 1135 for pointed cones, or NSWC reports for blunt cones. However, it is very cumbersome to have to interpolate manually in those documents for a given assigned task. Consequently, I decided to create my own utility codes to generate those solutions (for example, see page 90).

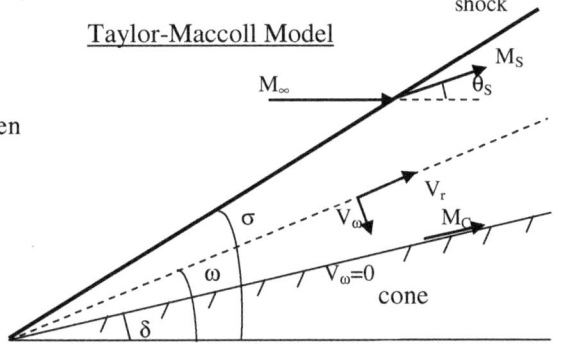

Taylor-Maccoll Model

I started with the Taylor-Maccoll solution scheme for pointed cones in supersonic flight from Ascher Shapiro's classic text book (1954). Flow is constant along conical rays emanating from the cone vertex. The potential-flow equations (inviscid and irrotational) are solved between the cone and the conical bow shock wave using a shooting (iterative) method (Ref 1). I named my resulting computer code **SCONE** (Ref 60).

The next step was to model flow over non-conical bodies. A very useful and simple method is to apply Newtonian Theory, which is formally valid only for high Mach numbers. However, this is not too limiting since the majority of a rocket's flight occurs at high Mach number. Solutions from SCONE showed that at high Mach numbers, the shock wave lay close to the surface of the cone. Consequently, the velocity behind the shock would be nearly parallel to the surface, and the component of free-stream momentum normal to the shock would be canceled at the surface, so must be zero behind the shock. The resulting pressure coefficient at a point on a surface at angle θ to the "freestream" is well known to be

$$C_p = 2\sin^2\theta \qquad \text{where} \quad \theta = \delta + \alpha\cos\phi \qquad \alpha = \text{angle of attack} \qquad (13)$$

However, this expression for C_p doesn't match the solution at the stagnation point of a blunt body ($\theta=90°$), where the gas must satisfy the jump relations across the detached normal shock. Consequently, the coefficient "2" in the above equation decreases as the stagnation point is approached, to 1.8 at the stagnation point for Mach numbers greater than 5.

I coded the solution for arbitrary wall shape as computer program **NEWTON** (Ref 61).

Since Newtonian theory applies to any surface canted at angle θ to the free-stream vector, it was easy to allow for the effect of angle of attack. By integrating the pressure coefficient over the entire vehicle surface, the pressure forces and moments can be determined.

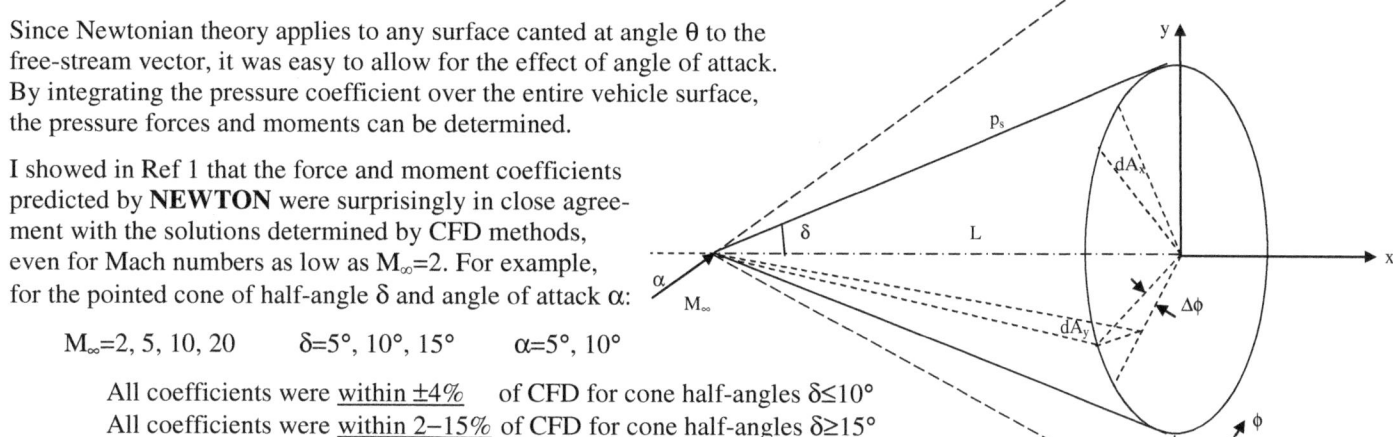

I showed in Ref 1 that the force and moment coefficients predicted by **NEWTON** were surprisingly in close agreement with the solutions determined by CFD methods, even for Mach numbers as low as $M_\infty = 2$. For example, for the pointed cone of half-angle δ and angle of attack α:

$M_\infty = 2, 5, 10, 20$ $\qquad \delta = 5°, 10°, 15°$ $\qquad \alpha = 5°, 10°$

All coefficients were <u>within ±4%</u> of CFD for cone half-angles $\delta \leq 10°$
All coefficients were <u>within 2–15%</u> of CFD for cone half-angles $\delta \geq 15°$

For blunt (spherical) noses, **NEWTON** predicted the axial force coefficient to within 5% of CFD predictions for Mach numbers between 3 and 20.

The shapes of vehicle nosetips are often defined by different parameters in different reports. In order to simplify the correspondence of those definitions, I solved, tabulated, and documented the mathematical relationships for blunt cones, tangent-ogives, and sphere-ogives as a function of which geometric parameters were specified (Ref 1). The geometric parameters are defined below on the upper half-plane:

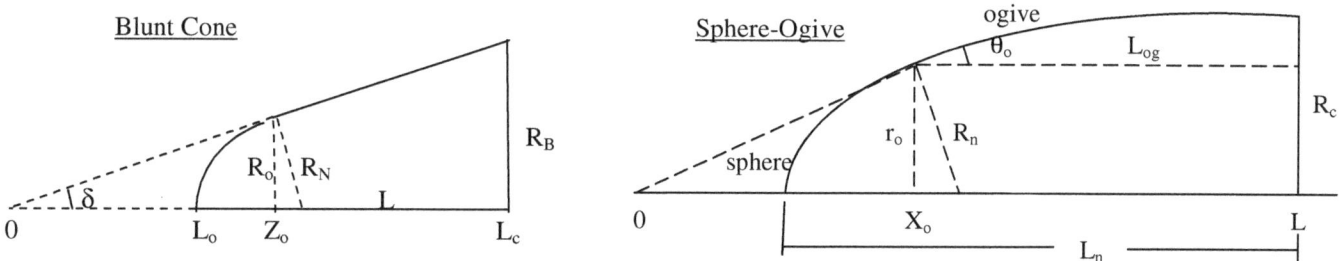

Level 2: CFD Solutions

CFD solutions are required for more-complex body shapes. Some of the vehicles for which I had to predict coefficients are shown below:

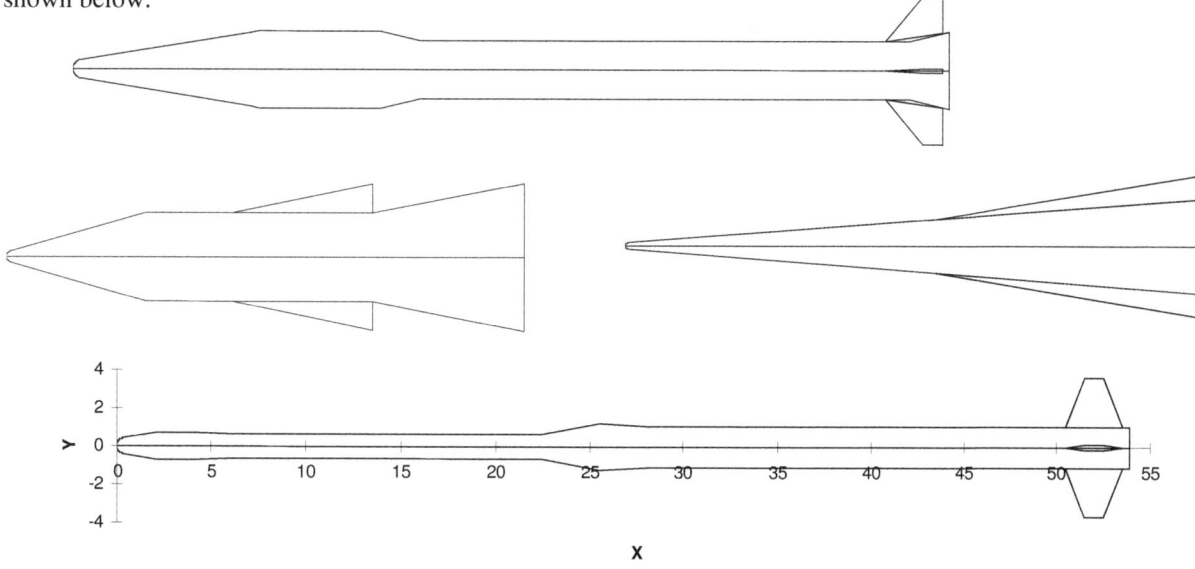

Three different CFD codes are summarized on the following pages: (1) a compact supersonic solver **ZEUS97** with built-in grid generator, (2) a blunt-nose subsonic/supersonic code **MN2IT**, and (3) a general subsonic/supersonic commercial CFD code **CFD-ACE**.

ZEUS97 for Vehicles with Fins

The CFD code ZEUS was developed in the 1980s and made user-friendly by me in 1997 (see page 64). It solves only super-sonic inviscid flow over finned or finless vehicles. Because the resulting governing equations are hyperbolic, they allow for spatial marching of the computations from a supersonic start plane just aft of the nose, then downstream to the tail of the vehicle. ZEUS creates its own computational grid as it marches. Its primary function at TRW was to predict the aerodynamic coefficients of the vehicle (axial force $C_A=F_A/F_\infty$, normal force $C_N=F_N/F_\infty$, normalized center of pressure location X_{CP}/L_{ref}; $F_\infty=Q_\infty A_{ref}$, $Q_\infty=\frac{1}{2}\gamma M_\infty^2 p_\infty$).

At the time, ZEUS was unique in being able easily to calculate the flow around fins.

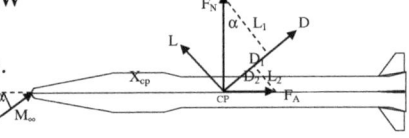

Flow properties across a supersonic start-plane are necessary to start the axial marching.
For a pointed body the start-plane is fully supersonic and easily calculated at zero angle of attack by the engineering methods discussed above. However, the start-plane is more complex at angle of attack. Consequently, a setup code INIT was provided with ZEUS to create the start-plane file for pointed noses; as noted earlier, I had to fix a significant error in INIT.

However, INIT was not appropriate for blunt-nosed vehicles. Consequently, I modified an existing blunt-nose flow code MN2IT to write the start-plane file in exactly the form expected by ZEUS (see next page). That was no easy task, since the MN2IT and ZEUS source codes were pretty unfriendly.

But, eventually I got it done.

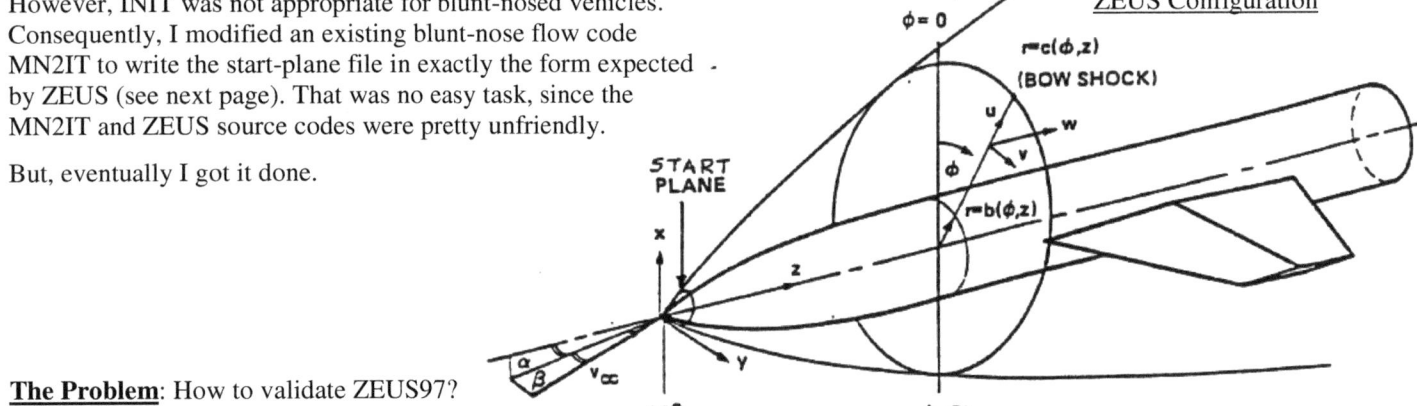

The Problem: How to validate ZEUS97?

My Solution:

(1) I compared its predictions to engineering solutions for pointed and blunt cylindrical bodies over a wide range of cone half-angles, bluntness ratios, and free-stream Mach numbers; agreement was generally excellent.

(2) I found test reports for a variety of vehicles in the TRW archives; configurations and measured data were from NASA/Langley, JPL, Naval Ordnance Test Station (NOTS), and Missile Ordnance Labs. I still have copies of the reports in my closet.

I typically ran multiple analysis codes for each vehicle configuration. For example, the finned ogive/cylinder shown on page 64 was tested at NASA by Trescot/Foster/Babb at Mach 2.86 and 4° angle of attack. The ZEUS97 input file is shown below:

```
ZEUS TEST CASE #12a: NASA/TRESCOT, T12 FINS
&INITS   ACH=2.86, PITCH=4.0, YAW=0.,  NA =36,    MA=36,
         ZS =1.0,  B=36*0.32184, BZ=36*0.30107, BPHI=36*0.,        &END    INIT input
&INPUT   IMOD=1, ZETAEND=30.000, ZTAIL=39.187, KEND=50000,                 Reference parameters
         XLREF=3.0, AREF=7.0686,                                   &END
&GEOMB   NOSE=2, FINS=.T.,                                                  Body Geometry
         ZLOC= 9.00, 37.586, 39.187,
         BODY= 1.50, 1.50  ,  1.383,                               &END    Fin Geometry
&GEOMF   ZLE=34.116,  SWEEPL=47.15,  SSPANL=3.75,    NFIN=4,
         ZTE=38.297,  SWEEPT= 0.0 ,  SSPANT=3.75,    THICK=0.0,    &END
&CONVT   IZNN=2, NAN=36, MAN=36, MAZN(1)=18,18,                    &END    CONVERT input
&OUTPUT  IDEBUG=0, IPRINT=10000000, IPLOT=2000000, JSP=80*0,       &END    Output Controls
------------------------------------------------------------------------
```

The predictions from ZEUS97 and three other codes that I ran were compared to the measured data:

Parameter	Measured	DATCOM	ZEUS97	CFD-ACE	NEWTON
$C_A(p_b=0)$	0.290	0.319*	0.244	0.244	0.312
C_N	0.500	0.581	0.514	0.460	0.447
C_M	3.953	4.535	4.120	3.594	3.616
X_{CP}/D	7.907	7.807	8.013	7.813	8.084

* Includes base drag and 0.020 due to friction

Note that ZEUS97 generated the best agreement with the measured data. Comparisons to measured data for other vehicles of similar configuration showed that ZEUS97 was typically as accurate or better than most of the other prediction codes in use at TRW.

MN2IT for Blunt Noses

The Problem: Flow over a blunt nose has a stagnation point where the gas velocity is zero. The flow accelerates outward from the stagnation point, but remains subsonic for some distance. For supersonic flight, flow in the subsonic shock layer eventually passes the sonic line where it becomes supersonic. Specifying the flow properties on a "start-plane" beyond the sonic line provides a supersonic flow code like ZEUS97 with the required initial conditions for marching the computations aftward along the vehicle.

My Solution: The inviscid computer code MN2IT (Ref 62) was obtained to provide that start-plane. The body shape in the meridional plane is defined by a general polynomial, and the resulting shock layer is subdivided into MMAX points along the surface and NMAX points normal to the surface. The flowfield is solved using an iterative "shock-fitting" solution like that for SCONE. It took a significant effort to identify the required variables in the source code to write to the start-plane in the format required by ZEUS97, but eventually that was accomplished. An example of flow over a blunt cone is shown below.

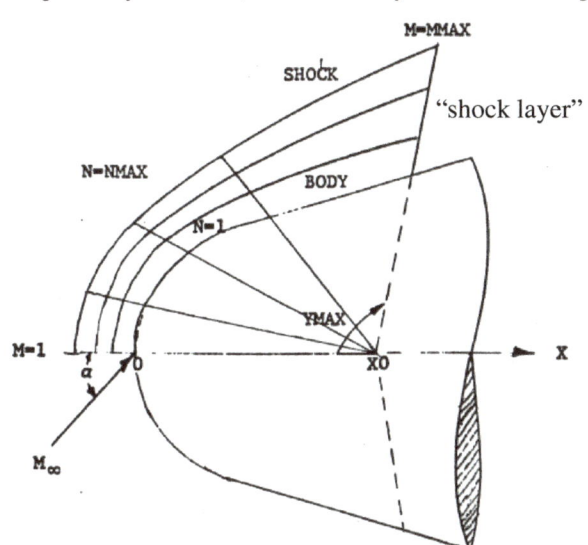

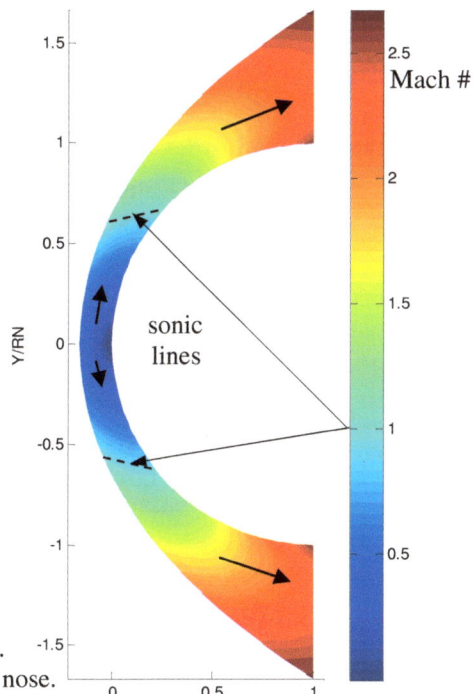

MN2IT: SUPERSONIC FLOW OVER AXISYMMETRIC NOSE
MN2IT TEST CASE #3b: BLUNT 10-DEG CONE, Rn/Rb=0.2, PITCH=0, Rn=3.526 NEW

In order to gain confidence in the code, the predicted pressure along the nose was compared to that predicted by Newtonian Theory. Over the range of Mach numbers from 2 to 10 and cone half-angles from 5 to 20°:

1) MN2IT was consistently only 3-5% lower than Modified Newtonian Theory, even at Mach 2.
2) Axial C_A changed less than 5% over the wide range of cone half-angles.
3) As expected, the blunt nose generated a much higher C_A than a pointed nose.

Component Buildup Codes AP98 and DATCOM07

Component buildup methods like DATCOM07 (Ref 63) and AP98 are useful for their ability to model aero coefficients in both subsonic and supersonic flow, and for exotic vehicle shapes like non-axisymmetric cross-sections. However, much artistry is required to run these codes reliably; for example, in AP98 it was unclear at the time whether biconic noses should be defined as a nose section plus a body section, or a double nose. Depending on the user's choice, different solutions were obtained. The original input format for AP98 was horrendous, so I converted it to Namelist format and ran it from a batch file. However, I never felt comfortable using AP98. Consequently, I generally avoided using it for standard axisymmetric body shapes.

The importance of plotting the body shape input to any aero prediction code was brought home forcefully to me when I made an input mistake to the Aero Prediction Code AP98. The original input required specification of reference length in feet, fin geometry in inches, and body coordinates in calibers. Confusing, especially when running multiple codes. In one simulation case I goofed up, but the body plot revealed that immediately by showing the fins flying about five feet behind the vehicle. That was a funny sight, but very educational:

AP98: VEHICLE GEOMETRY
PERSHING II: BLUNT BICONIC WITH 4 FINS, LREF=WRONG

I subsequently modified AP98 to require all input in a single set of units.

Commercial Code CFD-ACE

TRW bought a license for the commercial CFD code CFD-ACE from CFD-Research (Ref 59). I attended a training course in Huntsville Alabama, where I was taught how to use the GUI (Graphic User Interface) to create the input files for cases of interest.

The Problem: However, once I had created input files for several different cases, I found it was easier to clone one of those files and modify it directly for application to the next case. I felt it was inefficient and prone to user error to repeat setting up a whole new file using their GUI when most of the input was the same as the previous case.

My Solution: I built a suite of input files for my applications. Then whenever my supervisors asked me to rerun the existing case at new values of Mach number or pitch angle, it was easy to accomplish the rerun with confidence that the new and old runs would be consistent. I wrote a menu-driven DOS batch file RUNACE to operate CFD-ACE as well as their GUI did.

Another Problem: Then, one bad day, CFD-Research "upgraded" CFD-ACE such that the user could not run it by specifying an existing input file; instead it had to be run from the GUI. I objected, to no avail.

My Solution: So that I could directly rerun my many previous input files or their modifications, I was forced to continue to use my original version of the code. My desktop computer crashed several times over the years; luckily, Perry Daley at CFD-Research was able to find a copy of the executable for "my version", and to reinstall it on my computer. But he didn't think he would be able to do that for long. Eventually, the ability to run existing input files was returned using Python script, but I no longer used CFD-ACE (CFD++ had become the code of preference at TRW).

CFD-ACE had a convenient feature that simplified the treatment of fin aerodynamics. If the user constructed his 3D computational grid so that one grid plane was coincident with the plane of the fin, CFD-ACE allowed the flow properties on both sides of the fin to be modeled independently, thereby easily isolating the fin forces and moments.

I figured out how to generate a MAGG computational grid for a vehicle with two sets of fins for use in CFD-ACE:

MAGG...MULTIBLOCK ALGEBRAIC GRID GENERATOR
4-ZONE GRID FOR FLOW PAST VEHICLE B W/O STRAP-ONS (45x31+10x31+15x31+10x31 --> 77x31)

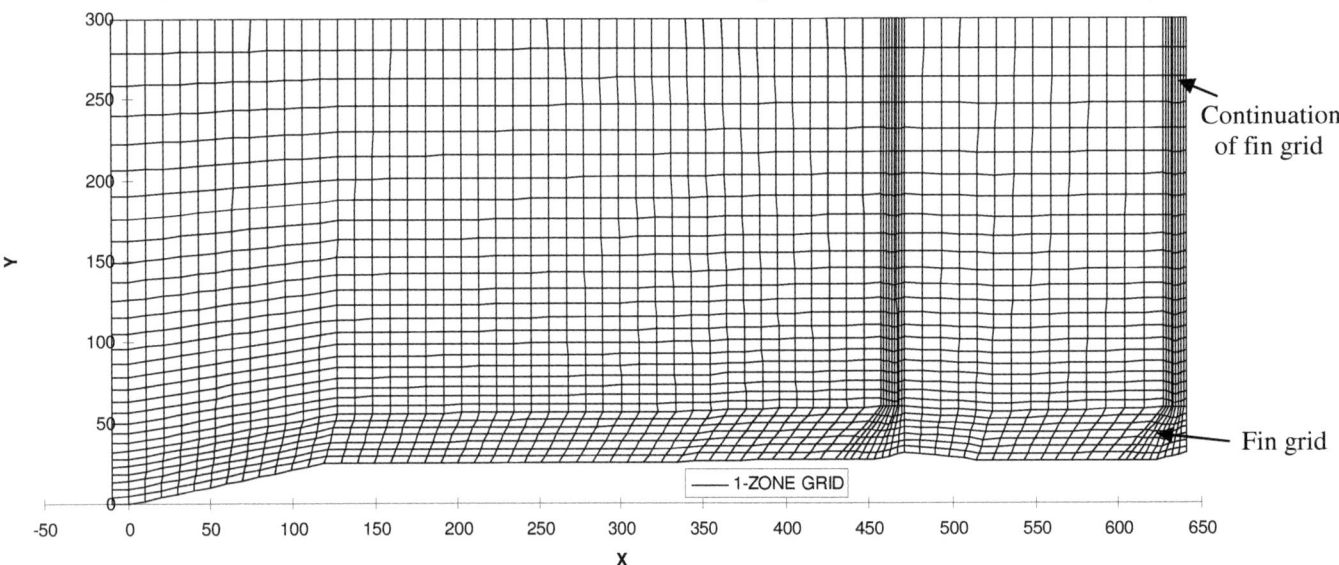

Agreement of the CFD-ACE-predicted aero coefficients with measured data was pretty fair for this vehicle and grid:

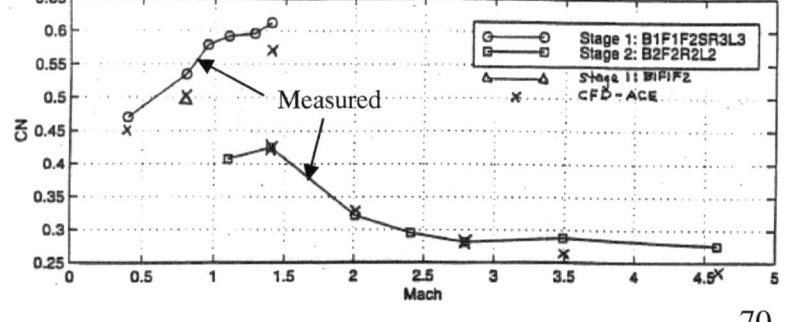

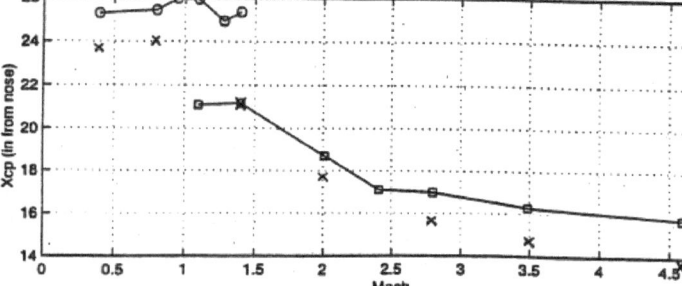

70

Nozzle Liner Ejection – Modeling with CFD-ACE

The Problem: The nozzle on a well-known motor has a graphite-phenolic liner (insulation) that extends from a location somewhat downstream of the nozzle throat to an area ratio of 10 where the nozzle extension is attached. This liner is ejected during many static firings and flights, but no earlier than 30 sec into motor burn. It is apparent from the many successful static firings and flights that the silica-phenolic backup insulation provides sufficient insulation during the remaining motor action time to prevent nozzle failure. However, the high failure rate of the graphite-phenolic liner was of concern.

Static firings of these motors revealed a temporary thrust drop of 0.5% due to liner ejection. However, the thrust eventually returned to nominal before the end of motor action time. I felt that this return to nominal occurred due to the burn-out of the forward-facing aft end of the vacated liner space (long notch).

My Solution: In order to simulate this phenomenon, I modeled the flowfield with and without liner ejection at a burn time of 30 seconds using commercial code CFD-ACE (Ref 59) on a 4-zone computational grid generated by my MAGG code. For the case of ejected liner, the zone 2 computational grid was extended into the vacated ("notch") region, making sure that the grid outside of the notch was unchanged to ensure consistency with the notch-less simulation.

The predicted Mach contours are plotted below, and show as expected that the aft-facing notch face emits an expansion fan while the forward-facing notch face generates a compression-wave fan. Both these flow disturbances disrupt the flowfield across the nozzle exit plane, and therefore the nozzle thrust.

An integration of the flow properties across the nozzle exit plane indeed predicted a loss in thrust of 0.5% compared to the flowfield without ejection. This agreement with test data created confidence in the model. Studies of <u>partial</u> liner ejection were also conducted for comparison to flight data.

CFD Modeling of the Effect of Nozzle Liner Ejection on the Flowfield

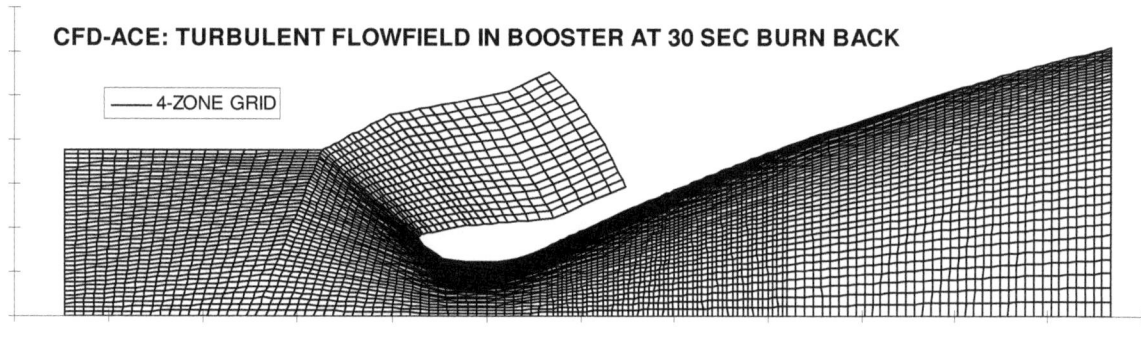

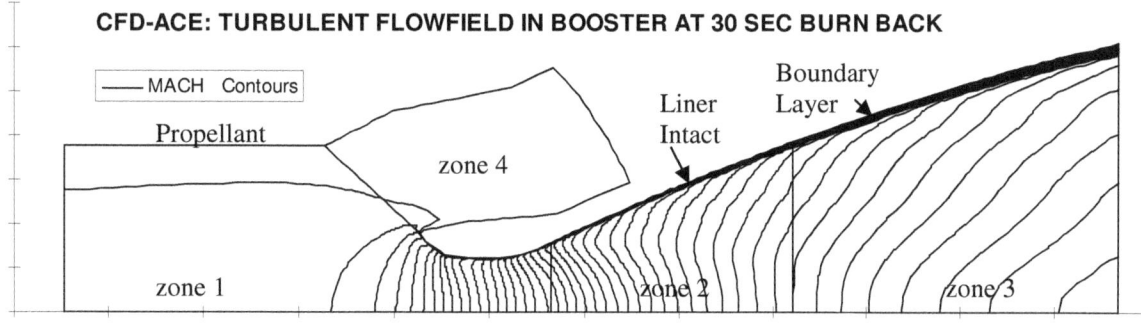

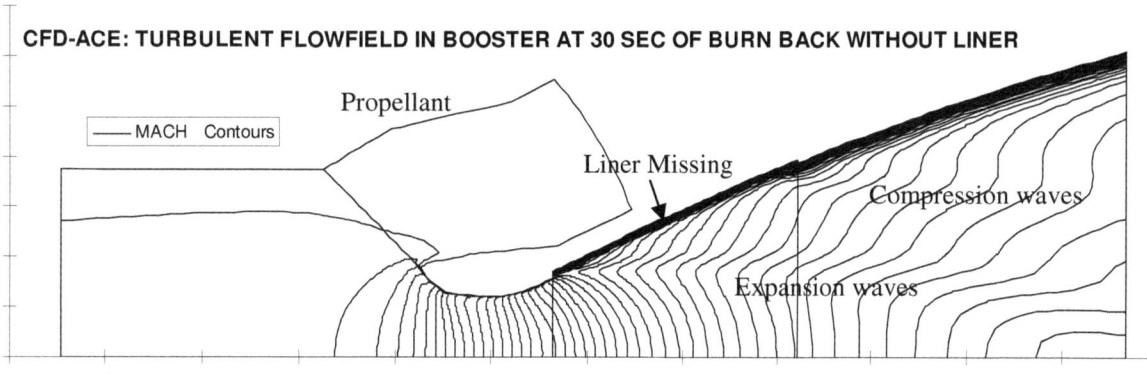

(5H) The Value of Closed-Form Solutions: A Lost Art?

The Problem: The flow in and around rockets is so complex that analysts need to have simple methods for understanding the physics, and even making first-order ("back of the envelope") quantitative estimates of the parameters involved. However, today's typical analysis environment often is:

- Run a commercial or unfriendly government-sponsored code on a 3GHz PC, Linux cluster, or supercomputer
- It generates output galore with pretty plots
- Is there a more efficient methodology?
- Can you believe the results?
  - Is code correct?
  - Is user operating it correctly?
  - Are input numerical controls reasonable?
- Sanity checks are necessary

Examples ("Horror Stories" that could have been avoided with my closed-form (C-F) solutions):

- Grain preheating: Thermal analyst chose wrong nodal spacing for CMA . . C-F solution exposed 30% error
- Log-normal integration: Analyst spent 3 pages of report verifying quadrature . . . C-F solution = $\exp(\lambda^2)$ = exact
- NASA-Lewis code: Near-vacuum plume thermochem bombed for $p \to 0$ . . . C-F solution independent of p
- SPP discrete Al_2O_3 sizes: Determined by quadrature   . . . exact C-F solution revealed quadrature error
- Chemical kinetics: High reaction rates stiff and expensive . . . approximate C-F solution provides alternative
- Reacting shock jump: CFD code predicted excessive temperature jump . . . C-F solution verified SPF-III jump

The Solution: There are many advantages of closed-form solutions:

- Quick solutions to simplified related problems provide insight and sanity checks
- Exposure of functional dependence on driving mechanisms
- Avoidance of discretization errors and smearing
- Sometimes generate stable solution when numerical codes go unstable
- Solutions easily modularized for inclusion as tests in numerical codes

For flow of <u>ideal gases</u>, the books of Ascher Shapiro (<u>The Dynamics and Thermodynamics of Compressible Flow</u>, 1953) and NACA Report 1135 (<u>Equations, Tables and Charts for Compressible Flow</u>, 1953) provided such solutions. I have automated the evaluation of these closed-form equations in my computer utility GASP (see Chapter 8).

Example: Culick's Neat Approximate Solution for a Cylindrical Combustion Chamber

For the non-ideal ("real") gases of the rocket environment, many engineers and scientists have derived and constructed additional closed-form solutions. For example, Prof. Fred Culick of Caltech derived a simple, elegant, and useful approximate solution for quasi-steady inviscid rotational incompressible flow inside a cylindrical solid-rocket combustion chamber with uniform propellant burn rate (injection velocity v_w):

$$U(x,r) = \frac{u}{v_w} = \pi X \cos Z \quad (14)$$

$$V(x,r) = \frac{v}{v_w} = -\frac{\sin Z}{Y} \quad (15)$$

$$\Psi(x,r) = X \sin Z \quad (16)$$

where $X = \frac{x}{y_w} \quad Y = \frac{y}{y_w} \quad Z = \frac{\pi}{2}Y^2$

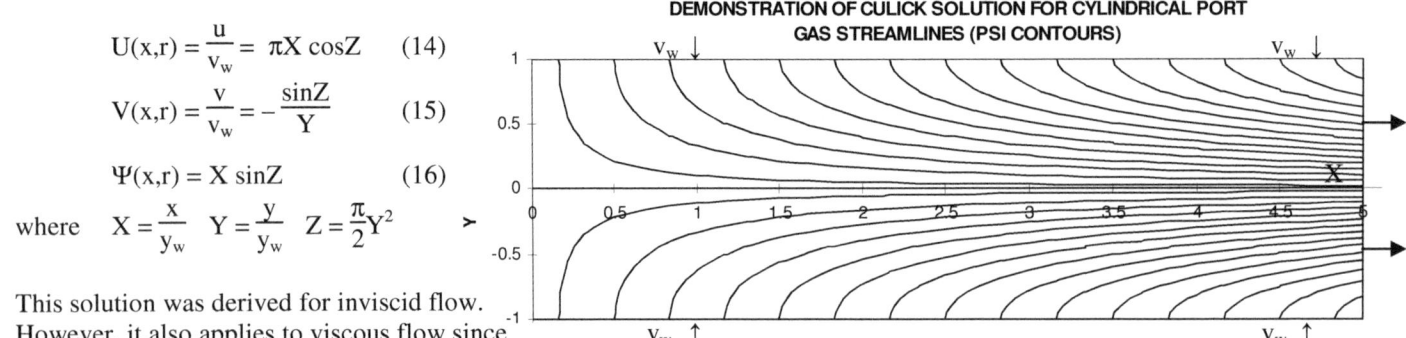

DEMONSTRATION OF CULICK SOLUTION FOR CYLINDRICAL PORT
GAS STREAMLINES (PSI CONTOURS)

This solution was derived for inviscid flow. However, it also applies to viscous flow since the inflow boundary is "no slip" (u_w=0). This has been confirmed experimentally by Dunlap, Willoughby, and Hermsen.

Numerous closed-form solutions of mine are presented in Refs 1 and 28; configurations for which I derived closed-form solutions for gas/solid heat exchange in rocket environments are shown graphically on the next several pages. Other useful closed-form solutions in gas dynamics have already been presented in this memoir: the "ballistician's equation" on page 20, my equilibrium thermochemistry model on page 34, Newtonian Theory on page 67, and my rarefied aero coefficients on page 74. For the serious student, the mathematical details of my closed-form solutions for chamber blowdown and for heating of planar surfaces are documented on pages 92-93.

Solutions to 1D heat conduction equation using Laplace Transformation and Convolution for constant heat flux to:

Solid Cylinder or Sphere   **Inner Wall of Hollow Thick Cylinder or Sphere**   **Inner Wall of Annular Cylinder**

Heat Soak of :

Thick Slab by Infinite Hot Gas Reservoir

Parallel Thick Slabs by Enclosed Hot Gas Reservoir

Heating of Thick Solid from Thin Flowing Gas Layer

COOLING OF THIN GAS FILM BY THICK SOLID

Some of these solutions are improved versions of those in the book by Carslaw and Jaeger; some they said couldn't be derived, but I did.

Sometimes I had to modify a colleague's solution. Phil Shadlesky derived a neat closed-form solution for the hemispherical emissivity ε of a radiative heat flux from a two-phase cylindrical medium (pVII-18 of Ref 1) in terms of albedo ω and optical thickness τ. His solution had $\phi=1$:

$$\varepsilon = \frac{4k\, I_1(k\tau)}{3I_0(k\tau)+2k\phi I_1(k\tau)} \qquad (17)$$

where

I_n = modified Bessel function of order n

$k = [3(1-\omega)]^{1/2}$

This solution was in excellent agreement with numerical solutions for a wide range of ω and τ. However, his solution was not correct ($\varepsilon>1$) for small ω at large τ.

I derived the function ϕ required analytically to force the correct asymptote at $\tau \to \infty$. The resulting solution was thus a team effort.

HEMISPHERICAL EMISSIVITY OF GAS/PARTICLE CLOUD
SALITA, AIAA 98-3965, 34TH JPC (CLEVELAND), 7/98

Aerodynamic Coefficients in Rarefied Flow: Closed-Form Solutions for Arbitrary Shapes

We had a project where we had to predict the axial and normal forces on vehicles and satellites flying at high altitudes where the air is rarefied. At high altitude (> 250 kft), the mean free path λ between collisions of air molecules among themselves is of the same order as with a vehicle of dimension L (i.e. the Knudsen number $\lambda/L = O(1)$). Consequently, the forces on the body are created by the transfer of momentum due to **individual collisions of molecules with the surface**.

The Problem: The Northop code FREEMOL that we were using was poorly documented and difficult to run.

My Solution: In order to help assure ourselves that we were running the code properly, I derived exact analytical solutions for axial C_A and normal C_N force coefficients (drag, lift) for the following shapes assuming isoenergetic specular reflections:

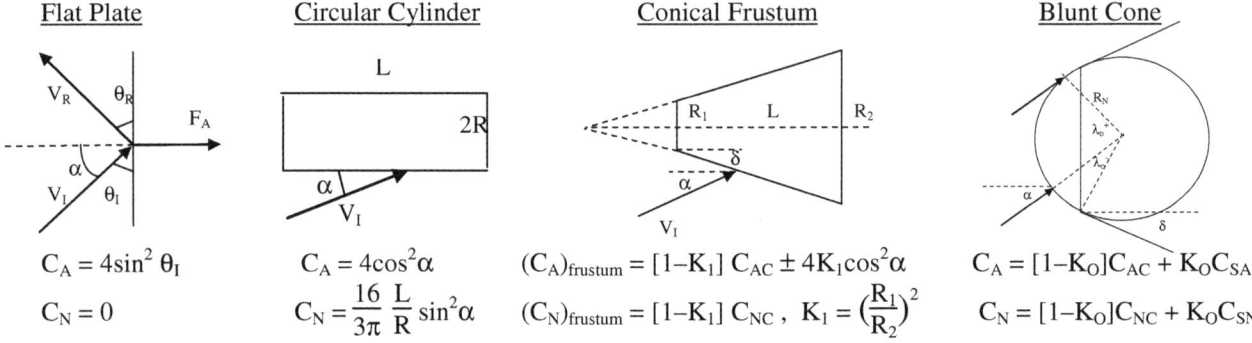

Flat Plate	Circular Cylinder	Conical Frustum	Blunt Cone
$C_A = 4\sin^2\theta_I$	$C_A = 4\cos^2\alpha$	$(C_A)_{frustum} = [1-K_1]\,C_{AC} \pm 4K_1\cos^2\alpha$	$C_A = [1-K_O]C_{AC} + K_O C_{SA}$
$C_N = 0$	$C_N = \dfrac{16}{3\pi}\dfrac{L}{R}\sin^2\alpha$	$(C_N)_{frustum} = [1-K_1]\,C_{NC}, \quad K_1 = \left(\dfrac{R_1}{R_2}\right)^2$	$C_N = [1-K_O]C_{NC} + K_O C_{SN}$

Note that the solution for the conical frustum is that for a pointed cone to radius R_2 minus that for a pointed cone to radius R_1. The blunt cone solution is that for a conical frustum, plus the solution for a spherical cap $C_{AS} = 2(1+\cos^2\lambda_0)$ where the angle λ_0 is shown for the domed cylinder below. Terms C_{AC}, C_{NC}, C_{SA}, and C_{SN} are documented in Ref 1, and $K_O=(R_N\cos\delta/R_2)^2$. Since many free-molecular applications involve tumbling vehicles, care had to be taken to account properly for negative pitch angles α, and the resulting shielding of the noses by the afterbodies.

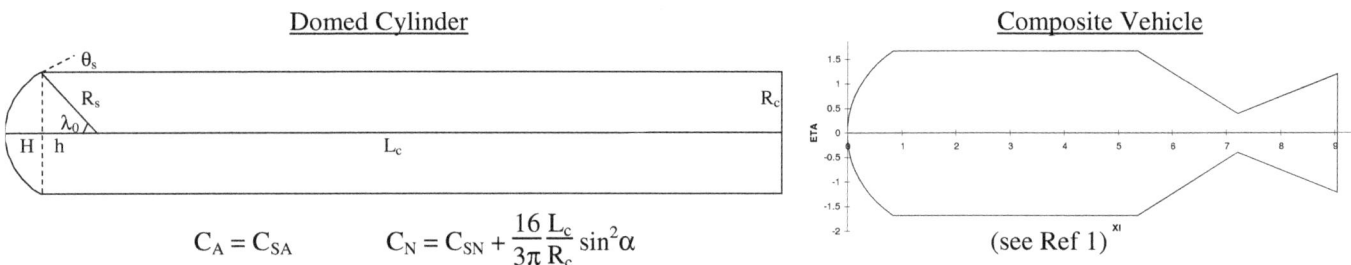

Domed Cylinder	Composite Vehicle
$C_A = C_{SA} \qquad C_N = C_{SN} + \dfrac{16}{3\pi}\dfrac{L_c}{R_c}\sin^2\alpha$	(see Ref 1)[xi]

Equations for the flat plate, circular cylinder, and sphere were previously available, but I have never seen those for the blunt cone, domed cylinder, or composite vehicle anywhere else. Those derivations were quite challenging because they had to account for tumbling or back-facing flight configurations, but their success was very rewarding.

Derivations of the axial and normal force coefficients are documented in Ref 1. Excellent agreement with the predictions from FREEMOL provided confidence in our use of FREEMOL, even for non-isoenergetic and diffuse reflection.

Fortran Compiler Conundrum: Lahey vs Microsoft (1998)

Microsoft issues security patches periodically. Suddenly in 1998, our Lahey90 Fortran compiler of choice no longer worked. (A compiler converts user language into machine-understood language.) It was determined that a DLL file in the latest security patch conflicted with a routine in the compiler. Lahey said that Microsoft had to change the DLL, but Microsoft said that Lahey was responsible to adapt. In the end, Lahey was the winner because their updated existing Lahey95 compiler avoided the problem. Consequently, all organizations using Lahey had to upgrade to Lahey95.

The problem was that Lahey95 was far stricter about coding rules than was Lahey90. Consequently, many codes that compiled fine using Lahey90 were now generating fatal errors. Luckily, most of the causes of these errors were easily fixed.

However, some programs required major modifications, in particular the Two-Dimensional Kinetics Code (**TDK**) in use throughout the rocket industry. I spent an entire weekend finding and fixing the sources of the errors. The principle problem was that character variables were embedded in many Common Blocks containing real variables, which was screwing up the storage bookkeeping. To fix the problem, I had to extract the character variables into separate Common Blocks. It took a lot of effort to ensure that all the Common Blocks were consistent among the many subroutines. However, by Sunday night TDK was compiling correctly.

Chapter 6: Northrop Grumman (2004-2009)

Northrop Grumman bought TRW in 2004. It then retained its Aerospace business, and sold its Automotive Division. My group at Northrop continued the TRW responsibility for managing the aging and reconditioning of the Minuteman ICBM fleet, and the silos in which they are stored, ready for launch.

Lead Pellet Problem

The Problem: During the casting of 23 Minuteman Stage 2 and 3 motors at UT/CSD, a door on an ingredient supply chute was sticking. A workman used a ball peen hammer to force the door open, but wasn't aware that the peen was leaking tiny lead pellets into the ingredients. High Energy CT scans subsequently revealed 391 "High-Density Inclusions" (the pellets) in the finished motors. The customer (USAF) refused to buy the contaminated motors unless it could be proven that the pellets would not affect the performance of the motors when fired. As overseer of the Minuteman fleet, Northrop Grumman was tasked to prove the sufficient quality of the motors.

Although the pellets were very small (with a maximum diameter estimated to be 2.2 mm) and unlikely to cause a problem, two steps were taken: (1) the motor with the largest number of pellets was test-fired, and (2) I was tasked to use my EVT model to predict whether the trajectories of the pellets would impinge on any nozzle surfaces during their expulsion from the burning motor. The tested motor showed no affect of the pellets, and my studies are summarized below.

Locations of Tiny Pellets in One Motor

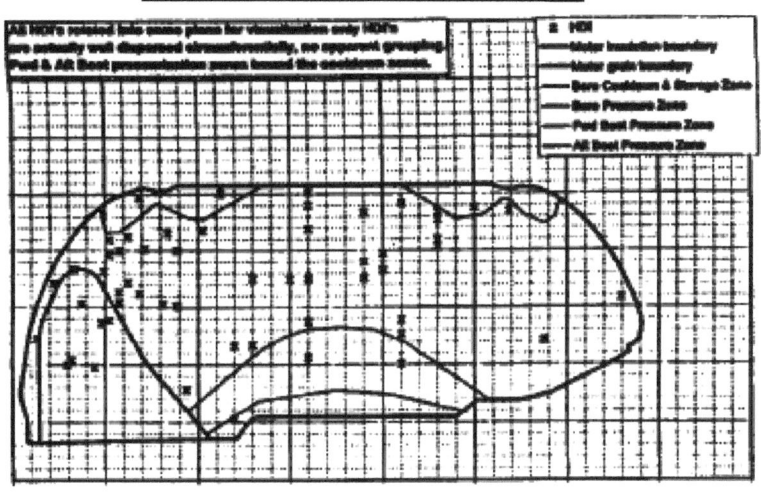

My Solution: When the propellant burn back reaches a solid lead pellet, the pellet melts and is blown into the flow where it begins to vaporize because of the low vaporization temperature of lead.

EVT was previously used to model the trajectories of aluminum oxide droplets expelled from the burning propellant surface. However, the droplet tracking module TD2P in EVT (see page 51) did not previously allow for droplet vaporization since the vaporization temperature of Al_2O_3 is much higher than the flame temperature of the propellant. Consequently, I had to modify TD2P to allow for pellet melting and subsequent droplet vaporization.

Burnback contours of the burning propellant were generated using SPP, and combined with my BALLIST code to identify the time and chamber pressure at which the burning surface reached each lead pellet. The pellet would then be ejected from the surface at this time, and the melting/vaporization process would begin. The gas flowfield and resulting trajectories of 80 of the most critical lead pellets were then calculated using EVT (Ref 64). This analysis predicted that all the pellets studied would have vaporized before reaching the nozzle in motors without axial acceleration. Incomplete vaporization was predicted for only a few of the pellets at 5 and 10 g's of axial flight acceleration, but their mass at impact was negligible.

Sample plots of gas streamlines and pellet trajectories with 5 axial g's are shown here; a <u>crosshatch</u> on any trajectory indicates the location where the pellet is predicted to have completely vaporized.

These EVT results helped convince the USAF that the lead pellets would not compromise the quality of the motors, so they bought the remaining 22 motors containing pellets.

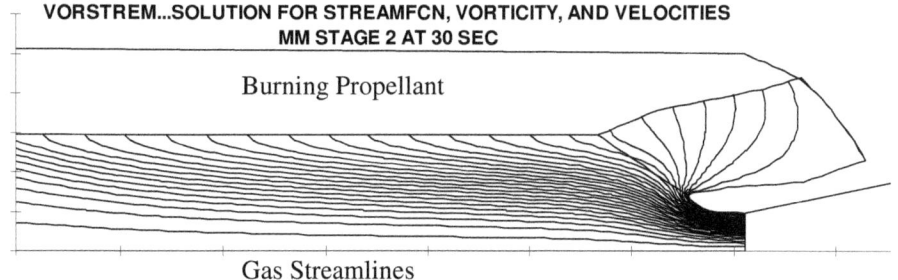

VORSTREM...SOLUTION FOR STREAMFCN, VORTICITY, AND VELOCITIES
MM STAGE 2 AT 30 SEC

Burning Propellant

Gas Streamlines

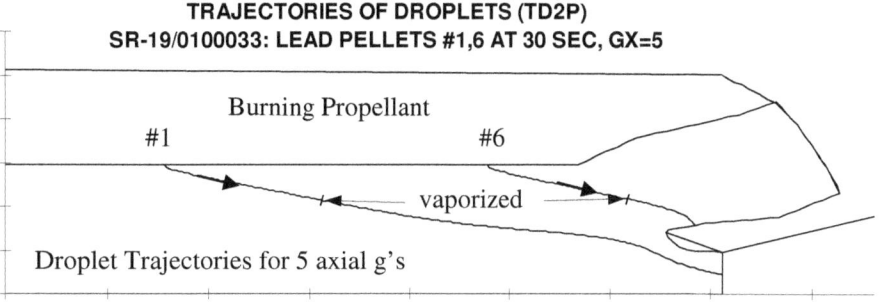

TRAJECTORIES OF DROPLETS (TD2P)
SR-19/0100033: LEAD PELLETS #1,6 AT 30 SEC, GX=5

Burning Propellant
#1   #6
vaporized

Droplet Trajectories for 5 axial g's

Modeling the Opening of a Minuteman Silo Cover (2005-2007)

The Problem: At first it might seem strange that the fluid-mechanical system for pulling open the cover ("closure") on Minuteman silos is very important (Criticality 1). But it becomes obvious when you realize that if the cover doesn't open, the ICBM fleet is useless. The DOD requirement is that the cover must slide fully-open within 6 seconds of the command that the gas-generators ignite to pressurize the piston that pulls the cable that pulls the cover sideways, against possibly a large pile of debris (e.g., snow and ice) blocking its path.

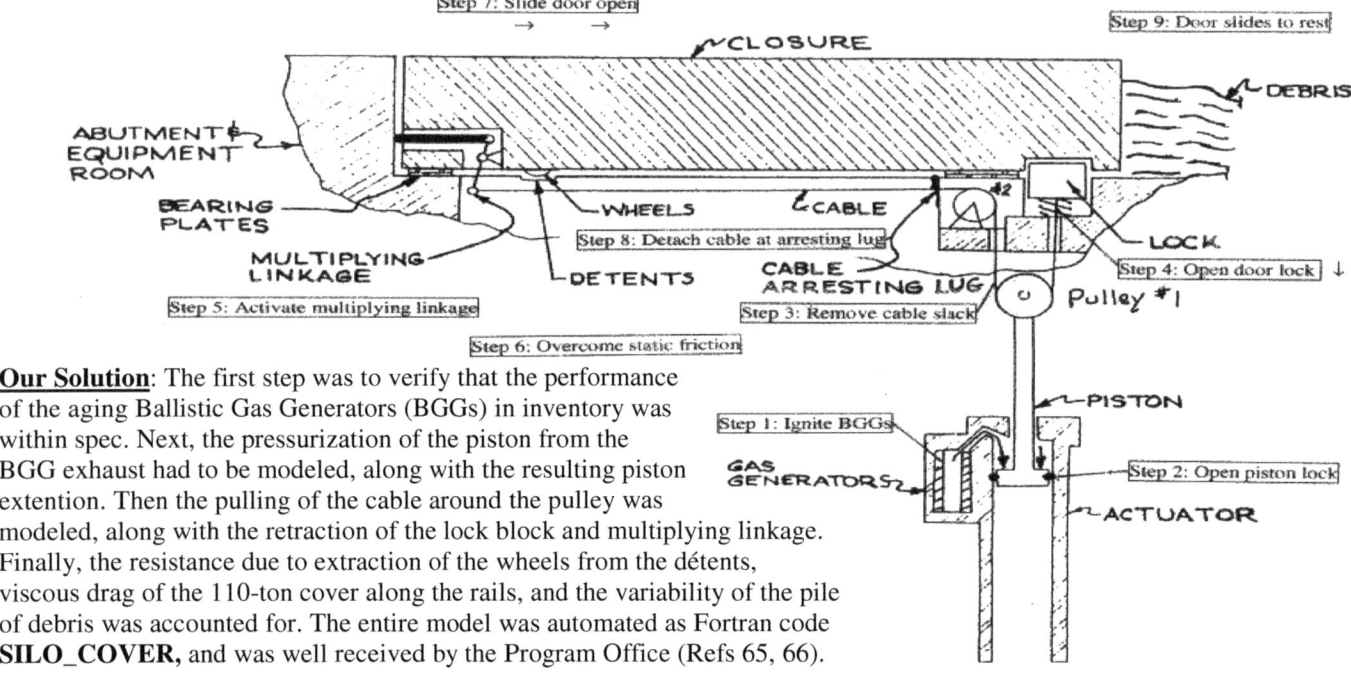

Our Solution: The first step was to verify that the performance of the aging Ballistic Gas Generators (BGGs) in inventory was within spec. Next, the pressurization of the piston from the BGG exhaust had to be modeled, along with the resulting piston extention. Then the pulling of the cable around the pulley was modeled, along with the retraction of the lock block and multiplying linkage. Finally, the resistance due to extraction of the wheels from the détents, viscous drag of the 110-ton cover along the rails, and the variability of the pile of debris was accounted for. The entire model was automated as Fortran code **SILO_COVER**, and was well received by the Program Office (Refs 65, 66).

Flash Fire in Minuteman Silo (2008)

My group was asked to model the short-duration flash fire in the Launch Equipment Room (LER) of a Minuteman silo that was caused by a faulty battery charger (Ref 2).

The LER in Minuteman silos contains 12 large water/acid batteries to provide power to the launch equipment in the event of an AC electrical power loss. In one year, the electrolysis process normally converts 6.0 gals of H_2O into 1096 ft^3 of H_2 at the cathode and 547 ft^3 of O_2 at the anode. Normal leakage out of the silo is sufficient to keep the H_2 concentration in the silo below the Lower Explosive Limit (LEL). However, a malfunction during an atmospheric electrical storm caused excessive electrolysis to occur over the following month, during which 12-15 gals of H_2O electrolyzed (as determined at subsequent LER servicing). Even this volume of H_2 would not exceed the LEL in the silo if uniformly diffused; however, a local region of high H_2 concentration apparently ignited from an equipment spark, causing a flash fire in the LER. Although the fire burned out quickly, this was a serious event in that the missile in the silo contained a nuclear warhead. Detailed CFD simulations of the air circulation inside the LER and launch tube have been conducted, and safety modifications were made.

Hot Gas Relief Valve (HGRV, 2007): A Neat Self-Correcting Device

A Liquid Injection Thrust Vector Control (LITVC) system injects jets of liquid into the nozzle flowfield from multiple ports in the nozzle wall. These programmed jets disturb the flowfield and therefore the pressure distribution over the nozzle interior surface, which provides control of the nozzle thrust angle. The pressure in the liquid supply tank of periodically increasing volume V_G must be kept constant as the liquid of volume V_L is ejected. The pressure in the system is provided by a constantly burning gas generator, but is maintained constant by a Hot Gas Relief Valve (HGRV, Ref 67) which vents excess gas.

The Problem: Cold flow testing of many HGRVs revealed several that did not perform within the spec window. A model was needed to identify the dominant performance factors and cause of the spec failures.

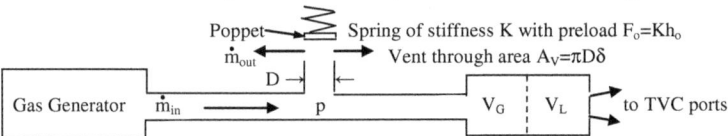

My Solution: I developed a simple model of HGRV operation. Initially a poppet of cross-sectional area $A_p = \pi D^2/4$ at the HGRV inlet is closed (h=0) and held in place by a spring of stiffness K whose preload $F_o = Kh_o$ was chosen to achieve a system pressure p_o. If the system pressure exceeds p_o, the poppet/spring is pushed upward a distance δ until $K[h_o+\delta]=pA_p$, and gas is vented out area A_V. The venting reduces the pressure, and any resistance to poppet/spring movement induces a larger pressure force to overcome the resistance. This self-correcting capability explains the high success rate of the design.

Pyrotechnic Ignition Anomalies

The Thiokol-produced **Standard Missile** (1980) had a magnesium/teflon pyrotechnic igniter that was lodged in a "donut" at mid bore. During static testing it was found that ignition was erratic (Ref 68). This problem was eliminated in two steps:

(1) It was determined that the aft end of the igniter case was rupturing too soon when the pyrotechnic pellets were ignited, so too much of the pyrotechnic energy was spewing axially toward the nozzle rather than into the propellant surface. Consequently, the aft end of the igniter case was reinforced to prevent premature rupture.

(2) I was asked to identify the optimum teflon content in the igniter. By running a series of thermochemical calculations and reviewing measured data for a range of teflon fractions and powder grind ratios, I determined that
 (a) early ignition is enhanced by using fine-grind teflon, and
 (b) a 50/50 mixture of magnesium and teflon will more than double the rate of grain preheating compared to a 60/40 formulation.

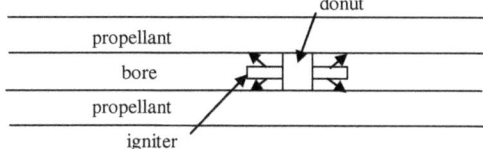

The resulting igniter led to a very repeatable and successful ignition process for the missile.

Hundreds of flights of the Bristol-produced **Black Brant** (2006, p80) high-altitude research rocket had operated successfully.

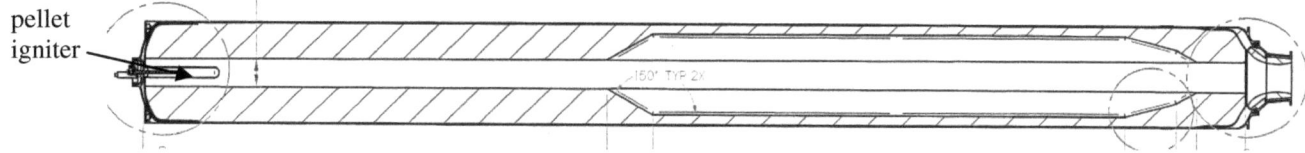

The Problem: However, a redesigned second-stage B/KNO$_3$ pellet-igniter failed to ignite the new reformulated HTPB main solid propellant during a test flight. Consequently, the igniter was to be redesigned again. The Northrop Grumman support group at Wallops Island was asked to help in the redesign effort, so they asked me to model the ignition phenomenon.

My Solution: I took the following steps:

(1) To predict ignition of the main propellant, the temperature of its surface at ignition must be known. Luckily, laser ignition data was available from NAWC for multiple values of constant heat flux Q. I was able to curve fit the time t_{ig} to self-sustaining ("go/no-go") ignition by $t_{ig}^{1/2}=17/Q^{0.92}$. However, for constant heat flux, I had previously derived the history of surface temperature T_s to be given by (repeated here on pages 92-93):

$$T_s(t) = T_o + \frac{Q}{k}\left(\frac{4\alpha t}{\pi}\right)^{1/2}$$
where α=thermal diffusivity $\quad$ k = thermal conductivity

Thus, at ignition the propellant surface temperature must be

$$T_{ig}(Q) = T_o + 17\left(\frac{4\alpha}{\pi}\right)^{1/2}\frac{Q^{0.08}}{k}$$
(almost independent of heat flux Q) $\hspace{2cm}$ (18)

(2) Recognizing that during the burnback duration t_{bb} of the igniter pellets, the mass flux is low early because of low chamber pressure, and low late because of small remaining pellet surface area, I assumed that the history of pellet mass flux would be roughly triangular with a peak value $\dot{m}_o=2m/t_{bb}$ where m was the total mass of all igniter pellets.

(3) I then ran the 1D unsteady Thiokol ignition code **SHARPIT** with a triangular igniter pulse using t_{bb}=150ms to simulate Bristol igniter tests conducted in an inert pipe of roughly the same length as the full-scale motor with the actual pellet weight m=375 gm. Agreement of the predicted history of pipe head-end pressure with the measured history was good up to 125 ms when the pipe unchoked (not accounted for in SHARPIT).

(4) SHARPIT was then rerun for the full-scale motor using this triangular igniter history but with a range of pellet weights; these simulations predicted that pellet weights less than 220 gms led to non-ignition.

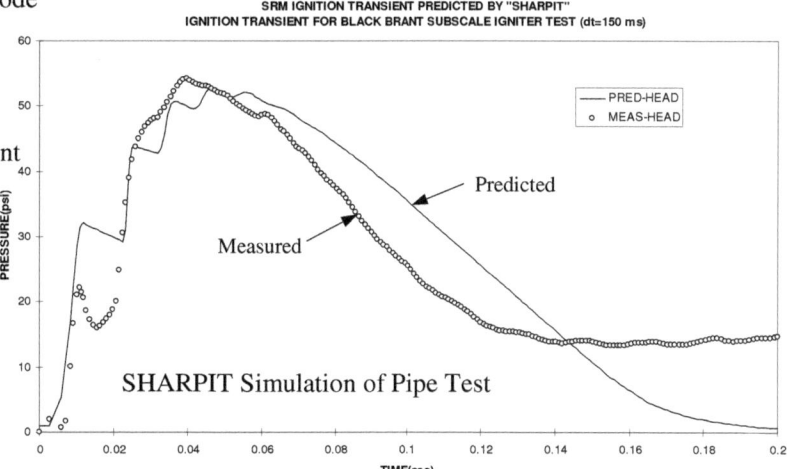

Since the pellet weight being used in the igniter was 255 gm, I felt that this weight was too close to the predicted non-ignition weight of 220 gms for us to be confident of successful ignition. I therefore recommended to Bristol that they increase their pellet weight (Ref 69). They did, and apparently haven't had a hang-fire of the motor again. See award page 80.

Launch Abort System for Orion Crew Module

The Problem: Manned vehicles for future U.S. space exploration will have a Launch Abort System (LAS) that can pull the Crew Module (containing the astronauts) to safety in the event of a motor failure during launch. Apollo had such a safety system, but Shuttle did not. When NASA issued a Request for Proposal (RFP) for an LAS for the Orion vehicle, Northrop Grumman decided to respond. My group was asked to design a "reverse-flow motor" that would generate sufficient thrust to pull the Crew Module away from the booster ahead of any detonation wave that might be generated by a failure of one of the booster motors below it.

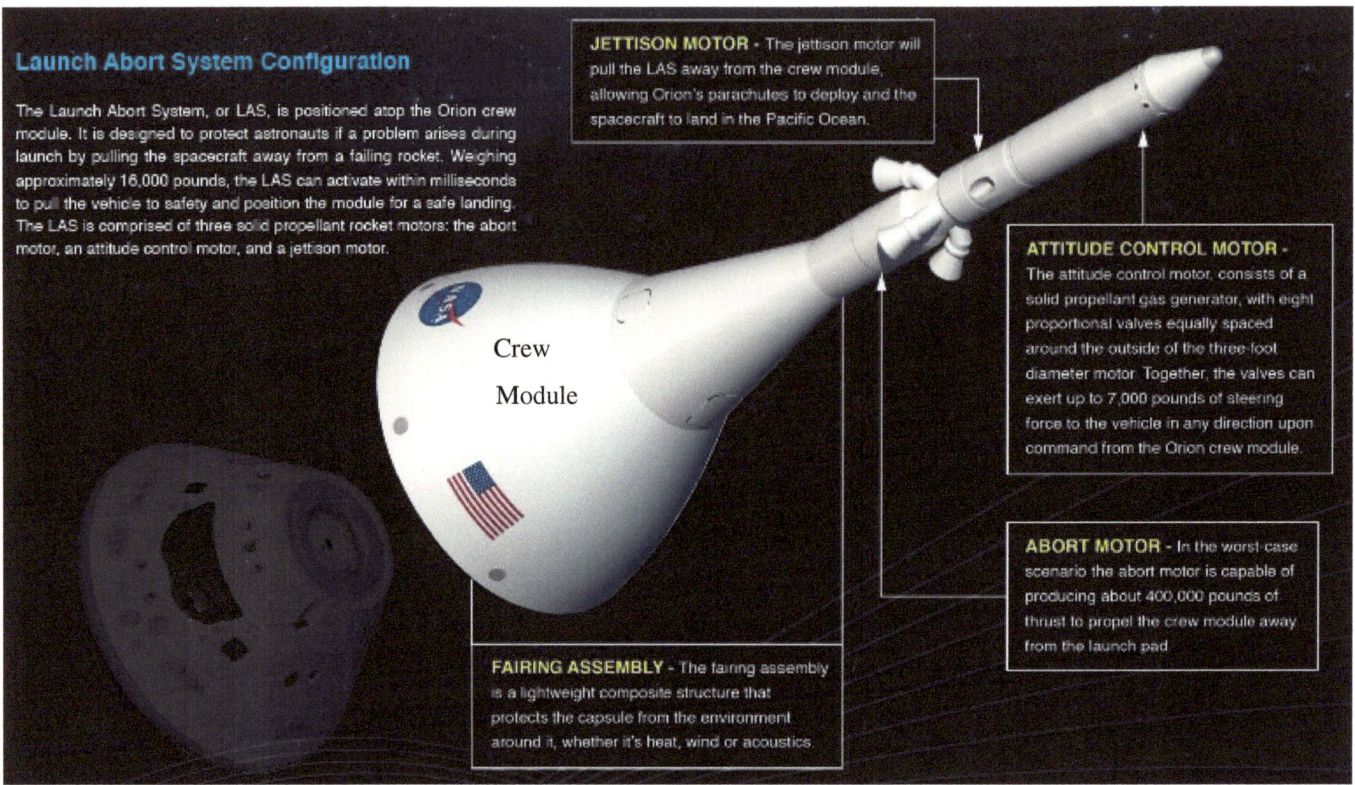

Our Solution: Northrop analysts Les Glatt and Kyu Hwang used commercial CFD code CFD++ to simulate its flowfield and the resulting heat flux to the Crew Module from the LAS exhaust plume. Given the heat flux, I estimated the required thickness of the CM thermal protection system (TPS) using my user-friendly version of the ablation code EXITS.

Plume Flowfield

Surface Heat Flux Footprint

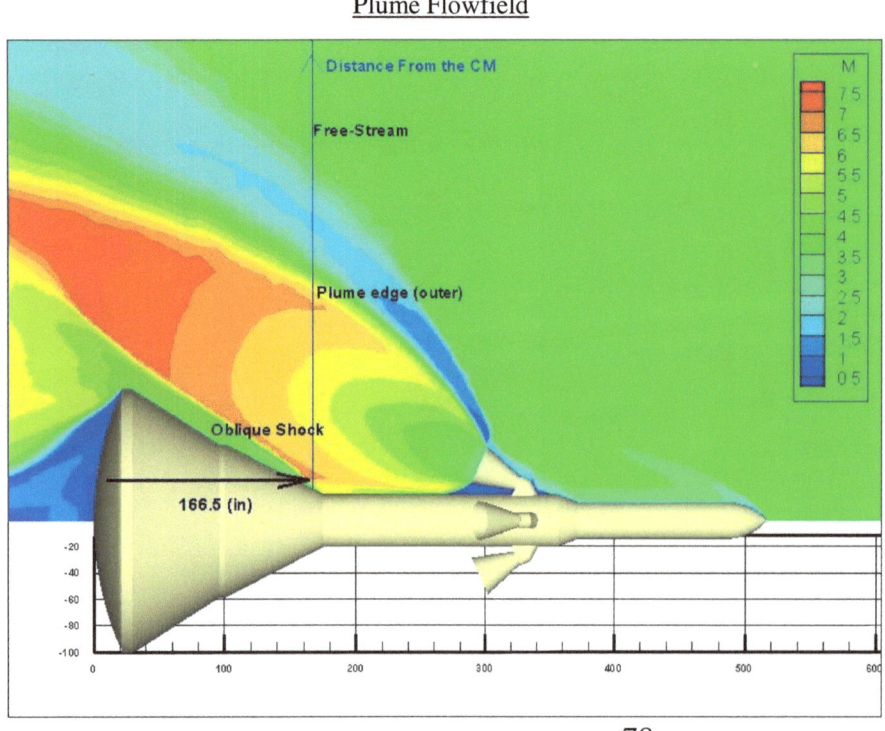

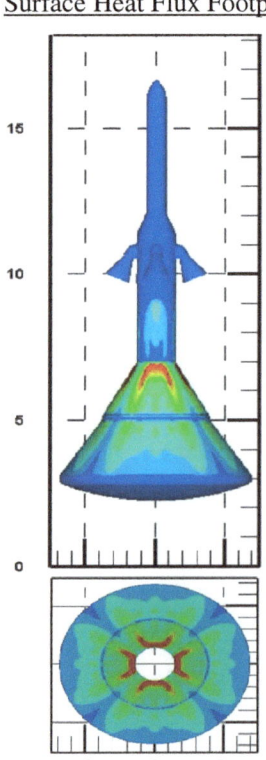

My Solution: My other task was to predict the propagation of the pressure and thermal waves caused by accidental detonation of the LOX/LH2 propellant of a lower stage liquid motor. To accomplish this ambitious task, I acquired the CTH detonation code from Sandia Corporation. This code was a very sophisticated but complex physical/mathematical/graphical model of the detonation event, with amazing graphics that I demonstrate below. As usual, I wrote a menu-driven DOS batch file to automate the multi-step procedure, which included (1) defining the special detonation chemistry, (2) predicting the propagation of pressure and thermal waves outward from the point of detonation, (3) generating the sequence of JPG flow-field pictures at specified increments of time, and (4) creating an AVI (movie) file from the sequence of JPG files (Ref 70).

CTH tracked the propagation of the missile materials after detonation: green is virgin structure (graphite epoxy), red represents the propellant combustion products, and blue is the steel from the propellant tanks:

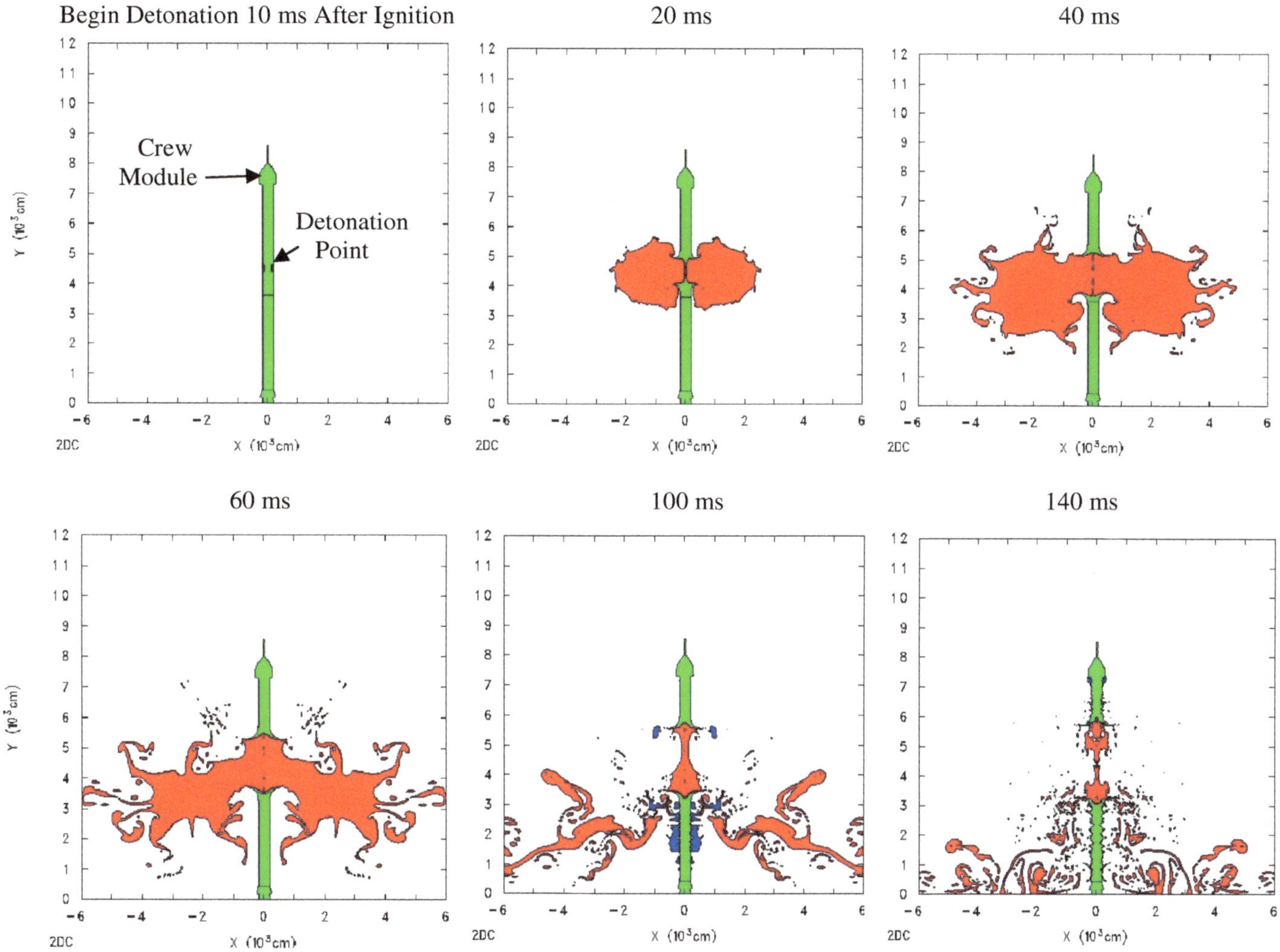

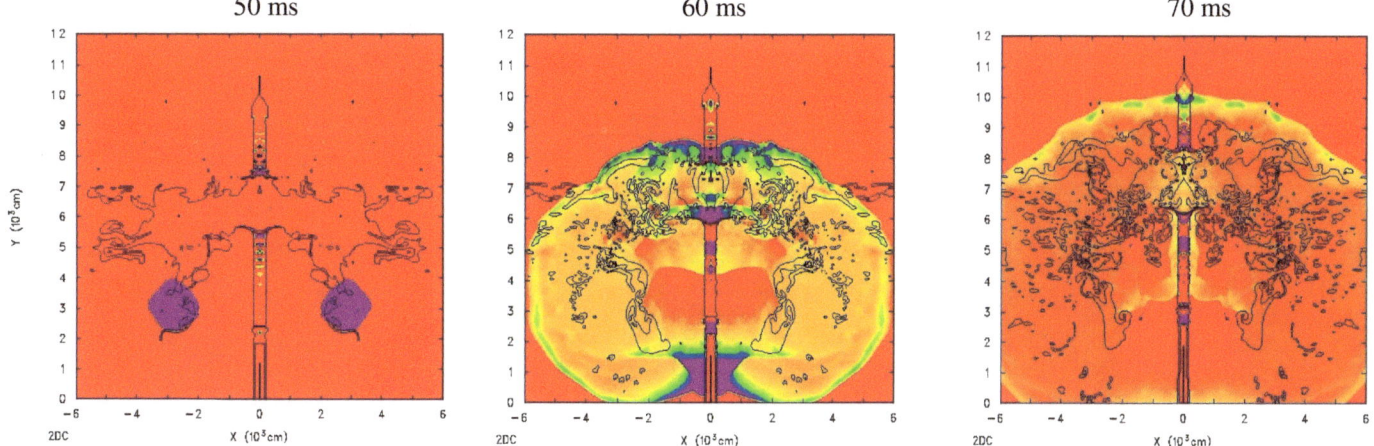

CTH also tracked the temperature bands;   The detonation thermal wave reached the Crew Module in only 70ms.

Ironically, Lockheed-Martin won the contract with a different LAS design, but changed to a design nearly identical to ours.

Black Brant

The **Black Brant** is a family of Canadian-designed sounding rockets built by Bristol Aerospace (a subsidiary of Magellan Aerospace) in Winnipeg, Manitoba. As of 2014, more than 1000 Black Brants of various versions have been launched since they were first produced in 1961, and the type remains one of the most popular sounding rockets ever built. They have been repeatedly used by the Canadian Space Agency and NASA. My participation in solving an ignition-failure problem in 2006 was summarized on page 77, and a resulting award is shown below:

The National Aeronautics and Space Administration

Presents the

Exceptional Achievement Award

to

Black Brant MK1 Return to Flight Team

For your contributions towards investigating and correcting anomalies associated with the new Black Brant MK1 rocket motor & restoring that system to fully operational status.

Dr. Mark Salita

Signed December Two Thousand Seven

Edward J. Weiler
Center Director, Goddard Space Flight Center

Automated Solution of Sudoku Puzzles (2012)

Like many others, I started solving Sudoku puzzles by hand. However, I soon realized that the solution thought process was always the same, and the challenge of automating the procedure would be more rewarding than solving them by hand.

Remember that the puzzle solution contains 81 single-digit numbers in a 9x9 matrix of locations; each location has 9 possible answers (1, …, 9), but no number may appear twice in any row or column or 3x3 sub-matrix. The solution process starts by viewing each of the 81 locations as a box, each with 9 possible answers. Input of the N initial specified numbers defines the single allowable number for those N boxes. The code then sweeps all rows, columns, and sub-matrices to eliminate all numbers in conflict with those N initial values. This creates additional boxes with only a single possible number. The sweeps are then repeated until all boxes contain only a single number satisfying the row, column, and sub-set requirement.

The resulting Fortran computer code SUDOKU was successful at solving automatically all puzzles of difficulty Levels 1-3 without any user intervention. However, in some puzzles of Level 4 (4 out of 7 tried), Level 5 (13 of 16 tried), and Level-6 (all 11 tried), user-intervention is required when some boxes contain at least two remaining possible numbers. In those cases the user must guess one or the other number in a box containing only two choices; if that choice doesn't complete the solution, then the user must repeat the process from the beginning and make the other choice. This process worked in 23 of the 25 puzzles requiring such user intervention. The remaining two (Level 6) cases required a second similar type of intervention. An example of a case requiring a single intervention is shown below:

LEVEL 6...GIVEN ARRAY OF 28 NUMBERS   51 BOXES SOLVED AFTER 5 ITERATIONS: CHOOSE 6 IN I=1, J=2   ALL BOXES SOLVED ... GOOD JOB!

Short Course

My department acquired a new hire who was an accomplished engineer, but had never worked on solid-propellant rockets. She was assigned several tasks, but soon complained that she had not received any appropriate training. I offered to create a "Short Course" to summarize the required technology that she needed to learn. The result was a viewgraph presentation that took days to present, and was subsequently given at both Hill AFB to Northrop employees and Air Force personnel, and at NOTS/China Lake. Unfortunately, the graphics left a lot to be desired, the material was limited primarily to the problems that I had worked, and the hand-outs were only on paper.

I was subsequently to extend this Short Course into a much more complete treatise on rocket technology, with additional content from other analysts in the industry, improved equation layout, and excellent graphics. The result was an electronic book in both text and bullet-chart forms (Ref 1). I used the bullet charts to teach from at the Technion in Israel, at Redstone Arsenal in Alabama, and at Orbital Launch Systems in Arizona. The cover is shown here on page 96.

An Ironic Story

I had an ironic experience only one week after 9/11, during a pre-scheduled trip to my TRW home office in San Bernardino. At the start of my return trip to Utah in the Ontario California airport, a TSA security agent asked me to hold out my hands so he could test them for gun powder residue. I chuckled at the effort, considering that I had a Secret Clearance and was part of the team that managed the U.S. ICBM fleet. I was no doubt less of a security risk than he was. I had an even bigger surprise when I noticed the name on the agent's badge: Mohamed.

Kinetic Energy Interceptor (KEI)

The KEI was a two-stage missile designed to intercept enemy ICBMs. The winning contract involved a team of companies (Raytheon, Northrop Grumman, Orbital, and ATK) which resulted in too many cooks in the kitchen. Communication channels were constrained by proprietary and security considerations, turf, and geography. Meetings were held in Arizona at Orbital in Chandler and Raytheon in Tucson.

There were five first-stage rocket motor tests planned to be carried out by Alliant Techsystems (ATK) in Promontory, Utah. The second test firing of a KEI _first stage_ rocket motor was conducted on June 14, 2007. The static firing included a full duration burn and a demonstration of the thrust vector control nozzle.

The project was classified, so there is little technical detail that I am allowed to provide here. The topics that I was tasked to address were

2) Help explain, model, and guide a fix of a "Ballistic Anomaly" observed in the first static test of stage 1.

3) Help explain and model an overprediction of delivered thrust in the first static test of stage 1.

4) Model the staging event with both STAGING and SALE3D and identify the amount of skirt venting required to minimize nozzle sideload during upper stage ignition.

5) Model the exhaust plumes at numerous times along each of several proposed KEI trajectories after launch.

The fourth test firing of the first-stage rocket motor was completed on November 13, 2008. The test demonstrated a successful operation of the first-stage rocket motor in its final flight configuration that was to be used during a summer 2009 flight test.

However, in May 2009, after three years of effort and just prior to the second flight test, the DOD canceled the contract. Remember, this was the Great Recession and the government decided that it could no longer afford the missile. Personally, I believe that MDA looked at the high cost of KEI compared to the much cheaper Israeli ABM Arrow 3 that was being co-developed with MDA and Raytheon, and decided that supporting Arrow 3 was more cost-effective. However, this was never acknowledged.

Hence, in 2009 Northrop Grumman needed to lay off some people. They asked for volunteers. I volunteered since I was already 66 and didn't want to delay my long-awaited desire to lecture at the Technion. The severance package was pretty minimal after 16 years tenure (two months pay), but I didn't complain, even when there was a miscommunication about the promised severance health coverage (they gave me less when they realized that I was already Medicare eligible).

I traveled to San Bernardino for my retirement party, and said goodbye to 32 years in the corporate rocket business. My parting words to my group were "**Communicate and brainstorm together. And document carefully the resulting thought processes, modeling, and results.**"

Luckily, there would be a lot more rocket challenges ahead in my life.

Chapter 7 – Rocket Science in "Retirement"

Teaching at the Israel Institute of Technology (Technion) (2009)

When the government canceled the Kinetic Energy Interceptor (KEI) program in 2009, Northrop Grumman asked for volunteers for a Reduction in Force (RIF). I was 66 years old at that point, and wanted to go teach my Israeli colleagues everything from my DOD-approved book before my mind started to atrophy. Consequently, I applied for the RIF. This would save positions for younger rocket scientists, and the severance pay would cover my expenses for several months. I notified my long-time Israeli friends of my availability, and was awarded a Lady Davis Fellowship to teach one semester to graduate students at the Technion (Israel Institute of Technology) in Haifa. Classes were to last 3 hours every Monday and Wednesday. The head of the Rocket Propulsion Center was Alon Gany, who had coauthored some important rocket journal articles with Len Caveny, with whom I am currently working to promote the UAV/BPI concept (see page 86). A small world.

My coordinator Benny Natan was very gracious at acclimating me on my arrival. I taught all classes in English, with which most of my students were very proficient (several had lived in the U.S. for years). Half the students were Russian émigrés. Ironically, the previous year's recipient of the Lady Davis Fellowship was Professor Merrill Beckstead of BYU. We knew each other well and had presented lectures in sequence previously, including at the Von Karman course described earlier. Although his fellowship had ended by the time I arrived, he and his wife had stayed in Haifa working for the LDS (Mormon) church. He had kept his office several doors from mine in the Rocket Propulsion Center, so we were able to interact periodically. He and his wife took me shopping more than once at the shuks and to a Druze village to buy gifts for my family.

I had an apartment on campus a short walk from the Rocket Propulsion Center, passing the Aero Building on the way. Most weekdays I went with one or more grad students to one of many campus restaurants for lunch, or with faculty members to a faculty dining room. Each had a wonderful selection of subsidized food, especially my favorite: salat chatzayleem (eggplant salad). For dinners, I often walked down to Ziv Square to have one of the best (so I was told) falafel sandwiches in Haifa; it was a meal in itself. A SuperSol grocery mart was across the street from the falalfel stand, so I often bought food there for my apartment, especially schnitzel, pita, and lots of hummus. Later in the evening, I walked the one block to the Recreation Center to use one of many treadmills. What with all the walking and treadmill time, I actually lost weight in Israel.

Incidently, as of 2015, 21% of Technion students are Israeli Arabs, and 48% of these are female. These are the same proportions of Israeli Arabs as in the total Israel population. A bus line connecting directly to Arab villages in the Galilee had a stop in front of my apartment. An example of equal opportunity in Israeli society!

Campus security was very interesting. Vehicles entering campus through the security gate would be stopped, and the driver asked to pop open the trunk for inspection. Even when the driver was the Director of the Propulsion Center or Administrative Dean. However, when either I or Merrill showed our faculty cards through the windshield, our vehicle was waved through without a glance. The first few times I came through on foot with a backpack full of groceries, the guards searched my pack. But after that they recognized me, and again barely gave me a glance. These are manifestations of the Israeli methodology for profiling and being sensitive to anything unusual. TSA take note!

I didn't want to spend months in Israel without playing my clarinet. Luckily, I was able before leaving for Israel to contact a professor at the Technion who was the flutist in a Wind Quartet. He told me that there would be a two-day music festival in Zichron Ya-akov while I was to be there. So I brought a clarinet, and we went to the festival. As a result, I finally got to play my favorite wind opus, the 13-instrument Dvorak Serenade Op.44 that previously I could only hear on recordings.

The final exam was scheduled to last three hours. However, I guess I made it too difficult, because no student had completed it in that time. So I let them continue, and continue, and continue. Eventually, the building custodian came and pled with me to terminate the exam so he could go home for dinner with his wife. I took pity on him, and ended up grading on a curve.

Ice Hockey in Israel

I couldn't spend an entire semester without playing ice hockey, or I'd be out of shape when I got home.. I knew there was skiing (on Mount Hermon) and ice hockey in Israel. So before I left for Israel, I checked the internet. Indeed there is an Israel National Ice Hockey Team that has been a member of the International Ice Hockey Federation since 1991. They even won the gold medal in the 2013 IIHF Ice Hockey World Championship Division II Group B playoffs. As a result, the team was promoted to Division II Group A. In 5 games played they were able to defeat China, Turkey, New Zealand and Bulgaria while suffering their only loss to Mexico. They were ranked 32nd in the world as of 2015. At first, it surprises many people that Israel even has a national team. But remember, more than 1.2 million people from the Former Soviet Union (FSU) have emigrated to Israel, including a half-million just in the 1990s. And you know how popular ice hockey is in Russia.

The main hockey rink is at the Canadian Center in Metulla, on the border with Lebanon. It was built by Canadian donors when Canada and Finland were UN Peacekeepers along the border in the 1980s, so they would have a place for recreation (there are also swimming pools, weight rooms, and even a bowling alley).

I was able to make contact with an Israeli from Toronto named Danny Spodek who ran a drop-in hockey group. He said I could play with them, and told me how to find the rink. "Follow the signs to Metulla. When you see the yellow gate, turn right and drive about a mile to the Canada Center". I asked "What if I miss the gate, since it will be night time?" He responded, "Then you will end up in Lebanon!" A chilling thought, but obviously there would be a road block at the border. Wouldn't there??

I played there several times. The group consisted of players from Toronto, Brooklyn, Finland, Russia, and Sweden. I had to converse with our Russian goalie in the few words of Russian that I remembered from my high school course.

Teaching at Redstone Arsenal (2010)

I was invited to present a week-long course from my book at Redstone Arsenal in Huntsville Alabama. I used the slides from my book. It was probably an overload for the students, but they needed extended-education credits for their job, so they were willing to endure the huge amount of material presented in such a short time. I probably should have scaled the scope down.

Consulting for Raytheon (2012)

I was contacted by Raytheon for help in solving an O-ring problem. I sent them my ORINGDEF code and plot packages, and instructed the assigned Raytheon engineer Sam Anderson how to use the codes. He was a quick learner, and soon had the code running successfully, including video displays of the O-ring seating using my MATLAB plot package.

Lectures at Orbital Launch Systems (2013)

I gave a free copy of my book/CD to Vince Allen at Orbital Launch Systems, which is located only 2 miles from my house in Chandler, Arizona. As a result, he recognized what I had to offer, and he paid me to lecture to his department on a number of topics of interest to them. I also suggested that since I lived so close to his facility, Orbital should retain me as a consultant. I could show up within minutes of a request to attend a meeting, with zero transportation or lodging cost to Orbital. That plan was pursued, including my registering in Arizona as a Limited Liability Corporation (LLC). It seemed like an ideal arrangement. However, at the last moment, Orbital management decided against any arrangement (they have a reputation for eschewing the hiring of consultants).

When people ask me why I don't charge for my book, I tell them that it is my legacy to the rocket industry. Besides, as with the lectures at Orbital, the book/CD sometimes gets me an opportunity to get remuneration in other ways. It's good PR.

Consulting for Spectral Sciences Inc. (2013-2015)

In 2001 I had been working at TRW with Mark Olmos to use the plume signature code SPURC. The code has many bells and whistles (exotic options), and is difficult to understand, especially if the user is not a specialist in signature analysis. It is not easy to learn a topic as sensitive as plume signature since much of its technical content is classified. Consequently, I decided to try to write a simplified computer program to learn the basics. I was given the names of two specialists: Bob Lyons at AFRL and Larry Bernstein at Spectral Sciences Inc. Larry was very gracious in walking me through the learning process, from which I created the program PLUMESIG. In the process, he became acquainted with what I knew about SRMs.

Consequently, when SSI won a contract years later (2013) that required combustion and flow knowledge of SRMs, Larry invited me to serve as consultant on the project. I was happy to help. The contract was to model the signature from post-burn SRM chuff (primarily ejected insulation and slag) or thrust termination (TT) plumes. We focused on Minuteman since that provided the most data, plus I had already modeled the internal flowfield at TRW for the IPIC contract, and at Northrop Grumman for the lead pellet problem. Metacomp CEO Sukumar Chakravarthy was also a team member and ran CFD++ simulations that I was subsequently told agreed well with my EVT simulations. Another validation of my EVT package.

Helping AEDC: Making the Russian Plume Codes NARJ/PRCJ User-Friendly (2011)

The Problem: In 2000, AEDC purchased from the Russians a set of axisymmetric reacting two-phase plume prediction codes: NARJ (Numerical Analysis of Real Jets) generates a nozzle-exit start-plane and plume flowfield for a wide range of external flows, and PRCJ (Prediction of Radiation Characteristics of Jets) calculates the radiation signature of the resulting hot plume. The good news is that the codes seem to run robustly and to generate predictions that have previously been validated against measured data. The bad news was that the code package was incredibly unwieldy: it was comprised of hundreds of subroutines that had to be merged into a single source code and executable for each run, and there were 45 different computational modules possible, each with its own input file. Bob Reed of AEDC asked me for help with the codes.

My Solution: I made this package of codes nearly trivial to operate and post-process (Ref 71). Six steps were taken:

1) Single source files for 29 of the computational modules were constructed from the myriad of required subroutines.
2) A title line was added to all input and output files to avoid user confusion (the original codes had no title lines).
3) An option to read the input in Namelist format was created that streamlined the input and avoided potential errors caused by the original column- and line-specific format. Many additional input options were created.
4) Executables were generated for each of the modified source modules, after fixing a few binary-file errors.
5) A post-processor NARJPP was created that converted the output files into scan-able forms that are also exactly the forms required by my EXCEL and MATLAB plot packages. Special modifications were also made for the plots, e.g. to normalize plume coordinates by the nozzle exit radius, to generate a start-plane file for SPF3, etc.
6) A menu-driven DOS batch file RUNNARJ was created that completely automated the operation of NARJ and PRCJ, including editing input, running the codes, browsing the output, and plotting the axial and radial flow profiles of gas, species, and particles, and property contours and color-bands for the entire axisymmetric flowfield.

Numerous demonstration and validation cases were documented (Ref 72). Some examples are shown below:

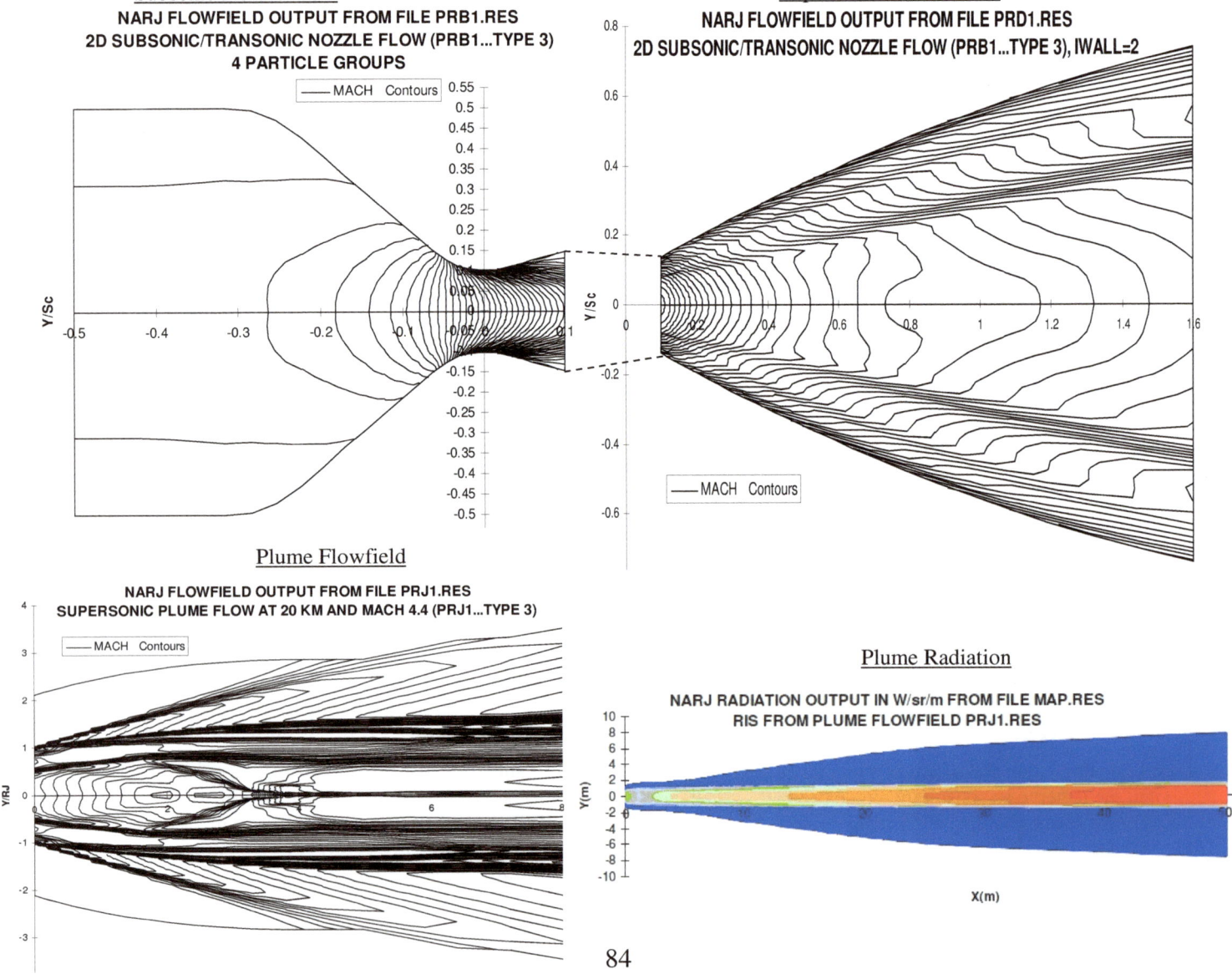

Subsonic Nozzle Flow

NARJ FLOWFIELD OUTPUT FROM FILE PRB1.RES
2D SUBSONIC/TRANSONIC NOZZLE FLOW (PRB1...TYPE 3)
4 PARTICLE GROUPS

Supersonic Nozzle Flow

NARJ FLOWFIELD OUTPUT FROM FILE PRD1.RES
2D SUBSONIC/TRANSONIC NOZZLE FLOW (PRB1...TYPE 3), IWALL=2

Plume Flowfield

NARJ FLOWFIELD OUTPUT FROM FILE PRJ1.RES
SUPERSONIC PLUME FLOW AT 20 KM AND MACH 4.4 (PRJ1...TYPE 3)

Plume Radiation

NARJ RADIATION OUTPUT IN W/sr/m FROM FILE MAP.RES
RIS FROM PLUME FLOWFIELD PRJ1.RES

Radiant Intensity of Rocket Exhaust Plumes (2014)

The Problem: Computations of spectral radiance $R(T,\lambda,L)$ of hot axisymmetric plumes depend on many parameters at each point in the flowfield (x,r): temperature T, wavelength λ, line-of-sight (LOS) path length L, pressure p, species mole fractions X_i, and "band-model" of absorption coefficient . Only three gaseous species (CO, CO_2, H_2O) plus two condensed species (soot, Al_2O_3) radiate significantly at infrared (IR) wavelengths over the range of temperatures of rocket exhaust plumes.

Four integrations are required to determine the total radiant intensity from the plume: (1) along each LOS through the plume at a single axial station, (2) across all LOS at that axial station, (3) over all axial stations comprising the entire length of the plume, and (4) over all wavelengths of the bandmodel.
These calculations are very expensive because the integration over wavelength must be carried out at every point in the flowfield, despite the fact that the bandmodel for each species doesn't change for a specified species, temperature, and pressure or particle size.

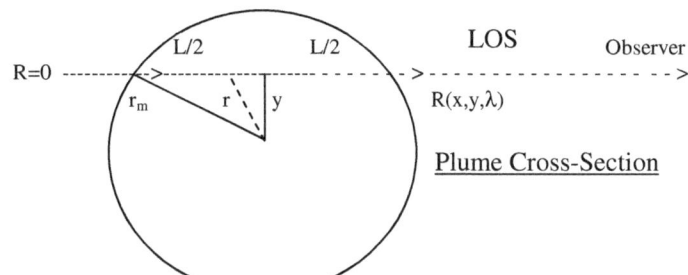

Plume Cross-Section

My Solution: I wondered if there was a way to uncouple the wavelength integration from the flowfield integrations. Then the bandmodels could be integrated over wavelength once and for all, and tables created for each of the five important species at each of its tabulated temperatures and particle sizes. Then at each point in the flowfield, interpolation in the tables at the local temperature and species mole fraction would determine the local emissivity at greatly reduced CPU cost. After studying the equation for radiant intensity at a single axial station across the plume, I determined that there was good news and bad news.

The bad news was that there was no way to uncouple the flow parameters from the radiative parameters for a general non-uniform flowfield, due to the exponential form of the radiant flux and the variation of temperature along each line of sight.

However, the good news was that if I assumed that the plume was radially uniform in temperature T, pressure p, and species mole fraction X_i at each axial station x, a simple closed form expression for the station radiant intensity R_{sta} could be derived for each individual species. For a plume of radius $r_m(x)$ containing a single chemical species at uniform conditions, the need for integrations 1 and 3 is eliminated. Integration (2) was eliminated by noting that the multiple LOS path lengths across the uniform cylindrical plume could be replaced by a single average value $L_{avg}=\pi r_m/2$. Luckily, the three parameters contributing to concentration combined into a single parameter $\beta=pX_iL_{avg}$. Integration (4) was carried out parametrically to generate tables of wavelength-averaged emissivity ε_{avg} versus temperature and concentration β. Then solutions were derived for R_{sta} and validated by comparison to predictions from the industry-standard numerical code SPURC for radially-uniform properties. Results were documented in Ref 73, and are summarized below:

(1) <u>Gaseous species</u>: $\boxed{R_{sta}(watts/cm)=2r_m\varepsilon_{avg}\sigma T^4}$ where emissivity $\varepsilon_{avg}(T,\beta,p)$ is tabulated for each species (dependence on p alone is only important at very low pressures where line-broadening occurs).

(2) <u>Particles</u>: $\boxed{R_{sta}=K_pNr_m^2\sigma T^4}$ is accurate for $K_pNr_m^2<5$, where N is the number density of particles of radius r_p, and $K_p(T,r_p)$ is the tabulated wavelength-averaged particle absorption cross-section in cm^2 for soot or Al_2O_3. Radiant intensity for multiple particle sizes can be linearly superposed, even though gas species can not be superposed.

Although no real plumes are radially uniform, the above equations (1) expose the dependence of station radiant intensity on plume radius, absorption coefficient, and particle radius and concentration, and (2) serve as a test of any numerical signature code. For example, when radially uniform flowfields were input to SPURC, the agreement of the above equations with the SPURC predictions was within 5% for the 500 cases run over a wide range of temperature, pressure, species concentration, and particle radius. A typical SPURC input file is listed below for a radially-uniform plume at $T=1800°K$, $p=0.1$ atm, radius $r_m=100$ cm (so $L_{avg}=157.08$ cm), CO mole fraction $X=0.1$ (so $\beta=1.5708$):

```
UNIFORM CYLINDRICAL PLUME OF LENGTH 50m, p=0.1, pLX=1.5708
     4
  R        T         P         CO
    0.        0.        0. 0.000E+00
     2
     2 0.000E+00      AXIAL STATION =    1
  0.000   1800.000 1.000E-01 1.000E-01
  100.0   1800.000 1.000E-01 1.000E-01
     2 5.000E+03      AXIAL STATION =    2
  0.000   1800.000 1.000E-01 1.000E-01
  100.0   1800.000 1.000E-01 1.000E-01
```

At these conditions, $\varepsilon_{avg}=0.00195$ so $R_{sta}=23.22$ w/cm^2 so total radiant intensity $R_{tot}=36.96$ kw/sr over the plume length of 5000cm. This value is only 0.7% lower than the SPURC value of 37.23 kw/sr.

UAV/BPI: Defense Against North Korean and Iranian ICBMs (2014-2016)

Strategy

The concept of "Boost Phase Intercept" (BPI) of ICBMs launched from "rogue" nations (e.g. North Korea and Iran) is well known to be the most assured way to intercept and defeat these missiles … hit them before they finish thrusting, i.e., before they begin their ballistic trajectory. Thus BPI occurs when their plumes are prominent and they are larger and slower and easier to intercept, before they can activate defensive countermeasures, and while the area from which they launch is much smaller than the area that must be protected.

UAV (Unmanned Aerial Vehicle, or Drone) platforms are much cheaper to achieve BPI than a seaborne flotilla or land-based system, and interceptor launch at 20 km altitude allows much higher intercept velocities due to greatly-reduced interceptor weight and aerodynamic drag. Unfortunately, during the 1980s and 1990s, UAVs did not have sufficient station-keeping time or payload to accomplish the BPI mission. Consequently, seaborne defense using Aegis missile cruisers or land-based Theater High Altitude Defense (THAAD) was developed, and the inertia of those large investments persists despite the asymmetry in cost of these surface-based systems compared to the cost of the enemy missiles. However, since then, the performance of UAVs has greatly increased, while the high costs associated with surface-based systems have become a budgetary issue. In addition, the threats from North Korea and Iran have become much greater. At the least, UAV/BPI would provide an additional layer of defense at minimal cost, and is an extension of the air-to-air combat mastered by the USAF and Navy.

Consequently, a former Thiokol colleague Len Caveny has been pushing the UAV BPI concept to industry and the Department of Defense (DOD). His idea is to station UAVs (e.g., Global Hawks or Avengers) at altitudes around 20 km (65 kft), at least a hundred kilometers from the rogue nation boundaries, and carrying two-stage hypervelocity interceptor missiles. As a former Science and Technology Director of the Strategic Defense Initiative Organization (SDIO) and the Ballistic Missile Defense Organization (BMDO), he has contacts high in the Defense Department. Nonetheless, he has as yet been unable to get them to pursue UAV BPI, and industry is loathe to pursue it without government support. Since Japan is more at risk from North Korean missiles due to their proximity, they should be more receptive to pursuing a UAV/BPI defense.

My Contribution

Len was using a trajectory program MISSILE-FLYOUT written at MIT by Geoff Forden to predict whether an intercept is possible for assumed properties of both the "threat" vehicle and the interceptor. He asked me to help. After studying FLYOUT, it was obvious the code was making some unnecessary approximations, and the rest of the model was inadequately documented (e.g. we never were able to identify its computational time step, and the treatment of earth rotation was never defined). Furthermore the code took way too long to run, for unknown reason.

Consequently, I wrote a 2-degree-of-freedom (2DOF) trajectory prediction code TRAJECT from scratch that is patterned after FLYOUT, but is more versatile and runs in seconds on a PC (Refs 74, 75). In addition, the predicted trajectories can be displayed immediately on Google Earth maps. For example, a plot of a notional North Korean Unha-2 (Taepodong-2) three-stage ICBM launched from North Korea and targeting LA is shown on the cover of these Memoirs; the boost phase is shown in red, while the intended trajectory is shown in white. The black traces are the missile trajectories that result from early thrust termination (for example due to interceptor impact) at either 100 sec (stage 1), 230 sec (mid-stage 2), or 340 sec (mid-stage 3). Note that if the interceptor simply terminates the missile thrust without destroying the warhead or remainder of the vehicle, the debris is predicted to hit Kamchatka Peninsula if intercepted at 230 sec, or near Anchorage Alaska if intercepted at 340 sec.

The figures below show (1) an intercept 230 sec into the Unha boost phase from an interceptor (orange trace) fired from a UAV at 54 kft stationed 200 km off the coast of North Korea, and (2) trajectories of three Taepodong-1 launches targeting Nagasaki, Tokyo, and Sapporo Japan (which were all shown in Ref 74 to be interceptable from a single UAV station):

Targeting and the Effect of Earth Rotation

The TRAJECT targeting model is pretty simple for a non-rotating spherical earth of radius R_o=6371 km. For a launch site at specified latitude ϕ_o and longitude λ_o, the angular distance $\Delta\theta$ and azimuth Ψ (0° is north, 90° is east) along a "great circle" to a target located at ϕ_t and λ_t is given on page 310 of Ref 76 (and rederived by me) as

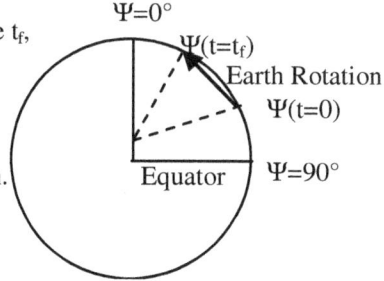

$$\cos(\Delta\theta) = \cos\phi_o\cos\phi_t\cos(\Delta\lambda) + \sin\phi_o\sin\phi_t \quad \rightarrow \quad \text{Required range } S_{req} = R_o\Delta\theta \quad (19)$$

$$\cos(\Psi) = \frac{\sin\phi_t - \sin\phi_o\cos(\Delta\theta)}{\cos\phi_o\sin(\Delta\theta)} \quad \rightarrow \quad \text{Required launch azimuth } \Psi_{req} \quad (20)$$

where $\Delta\lambda=\lambda_t-\lambda_o$. The equations of vehicle motion determine its history of altitude $y(t)=y_o+\int v\,dt$ and swept angle $\Delta\theta(t)=\int[u/(R_o+y)]dt$ given the current vehicle mass $m(t)$, and net thrust and aerodynamic forces F_x in the horizontal direction and F_y in the vertical direction. The thrust at launch is pointed at angle η_o relative to the ground (90° for ICBM vertical launch, 0° for UAV horizontal launch). This angle is decreased at rate $\dot\eta$ for vertical launches (or increased for horizontal launches) until a specified "loft angle σ" is achieved, after which η remains constant until burnout. In order to hit the target, the loft angle is iterated until the resulting calculated range $R_o\Delta\theta$ equals the required distance S_{req}.

Note that the above equations are independent of altitude; the altitude y_o of the launch site can be sea level or at the UAV location, and the altitude y_t of the target can be the surface of the earth or the ICBM intercept location.

The accounting for <u>earth rotation</u> is tricky. There are two frames of reference for this problem: an observer rotating with the earth, or a fixed ("inertial") observer looking down on the rotating earth. The earth rotates at a rate ω=361°/day=0.2507°/min, or 1524 ft/sec at the equator. During a missile flight of duration t_f from launch site to target, the earth will have rotated an angle ωt_f. Equations (19,20) for $\Delta\theta$ and Ψ still apply, but for $\Delta\lambda=\lambda_t-\lambda_o+\omega t_f$. Since these equations model the earth as rotating, then the observer must be inertial, so he sees a missile launch that includes the eastward component of earth rotation velocity (e.g. 1524 ft/sec at the equator). However, I used the Bate/Mueller/White model proposed in Ref 76 that does not add the component of earth rotation velocity to the surface-launched missile until it has achieved burnout (end of boost phase), based on the argument that that is when the missile is free of the earth's rotating atmosphere.

There can be several different definitions of range, so care must be used when interpreting other documents. The ground distance from launch site to target is the same whether the earth is rotating or not. However, missile "range" $R_o\Delta\theta$ on a rotating earth is the distance from the launch site at time t=0 to the target (or splashdown) location after flight time t_f.

The flight time is identified iteratively:
  (1) iterate on loft angle to hit the target for a non-rotating earth and note the flight time t_f,
  (2) update $\Delta\lambda=\lambda_t-\lambda_o+\omega t_f$ and the resulting required range angle $\Delta\theta$ and azimuth Ψ,
  (3) iterate on loft angle to achieve the new required range $R_o\Delta\theta$, and
  (4) keep repeating steps 2 and 3 until azimuth Ψ no longer varies.

This final launch azimuth is the value needed to "lead" the target during its earthly rotation.

For the Unha-2 attack on Los Angeles, TRAJECT predicts the following parameters to hit the "aim point":

Parameter	Non-Rotating	Rotating	
Range (km)	9579.2	10335.8	
y_{max} (km)	2774.3	2933.6	(y = altitude = r − R_o)
Flight time (sec)	2886.5	3065.1	
Azimuth (deg)	47.69	39.55	
Loft (deg)	50.72	50.96	

Consulting for the Univerity of Texas/Austin (2016)

A former colleague Dr. Joseph Koo from Thiokol, but now a professor at the University of Texas (UT) in Austin, asked me to support a Small Business Independent Research (SBIR) proposal to NASA JSC. It involved UT's use of their Inductively Coupled Plasma (ICP) Torch to measure heating and ablation of insulation materials intended to protect the Orion capsule during earth reentry. Consequently, I supplied UT with my plot package and copies of my user friendly versions of CET93, ACE and CMA92, and helped UT get them running. In addition, my version of CET93 contains Victor's simplified plume model; I showed UT how to run it to model the hot plume flowfield from their oxyacetylene torch.

UT also runs the widely-used FIAT ablation code from NASA AMES. However, I was astounded at how user-unfriendly the original version is. Consequently, I created a pre- and post-processor FIATPP and menu-driven DOS batch file RUNFIAT that makes the codes much easier to use. If UT wins any torch contracts, I will have many more tasks. Here's hoping!

Summary of My "Lessons Learned" in the Rocket Industry (2011)

As my legacy to the rocket industry, I documented 20 of my most important "**Lessons Learned**" in the award-winning AIAA conference paper "Unanticipated Problems and Misunderstood Phenomena in and Around Solid Rockets" (Ref 2):

Unanticipated Problems

1) ET-induced cooling of O-rings, coupled with wind shear at altitude
2) Wind-shear induced ET slosh, plus enhancement maneuver
3) Shroud-induced heating of aft dome of booster
4) Excessive overhang of low-modulus propellant at intersegment slot
5) Insulation debond in a supposedly unflawed motor
6) Anomalous lead pellets cast into solid propellant
7) Frequent ejection of liner in a booster nozzle
8) Excessive H_2 generation from backup batteries in Minuteman silo
9) Wet propellant during first steam launch of Peacekeeper missile
10) Overly-thick slot inhibitors in Ariane booster yields huge stubs
11) Accidental cold soak of missiles during storage

Unanticipated Consequence

O-ring failure and Challenger disaster
ET foam ejection and Columbia disaster
Failure of first Maxus flight
Detonation of first SRMU test motor
Test failure of baseline SBAS motor
Demise of UTC/CSD
Acceleration spike and temporary thrust loss
Flash fire in silo
Surprising ignition delay in canister launch
Severe pressure oscillations and high slag
Out-of-spec motors, non buy-off

Misunderstood Phenomena

1) Ignition overpressure (IOP) and wave reflection on launch pad
2) Effect of afterburning on peak ignition overpressure in silos
3) Nozzle side load during motor startup
4) Hoop strain-gage data during motor ignition
5) Looping of hoop strain vs chamber pressure
6) Dominance of groove wall on O-ring deformation
7) Misuse of inviscid flow code to model booster slot pressure
8) Erroneous understanding of Al_2O_3 slag accumulation
9) Use of closed-form manipulation of log-normal distributions

Consequence

Nearly snapped ailerons off STS-1 Orbiter
Importance of silo depth
Random direction and peak side load
Exposes igniter shock propagation
Exposes viscoelastic propellant behavior
Allows simple analytical O-ring model
Temporary stop of Shuttle booster program
Erroneous modeling of phenomenon
Allows efficient and accurate calculations

AIAA

The World's Forum for Aerospace Leadership

BEST PAPER
CERTIFICATE OF MERIT

For the purpose of promoting technical and scientific excellence

Presented To

Mark Salita

For the outstanding paper in Solid Rockets
"Unanticipated Problems and Misunderstood Phenomena in and Around Solid Rockets"

Presented At
47th AIAA/ASME/SAE/ASEE
Joint Propulsion Conference
August 2011

Basil Hassan
VP Technical Activities

Chapter 8 – Useful Tools

SUPERBAT

In 2007, knowing that I was going to retire soon, I thought it necessary for me to be sure to leave all of my computer codes and documents as accessible as possible for my Northrop colleagues. To accomplish that, I enlisted the services of a Northrop colleague who knew how to code a Microsoft Data Base (MDB) in Microsoft Access, and we constructed a GUI, which I named SUPERBAT. What this code does is serve as a single entry point to all of my codes and documents. Once the category and code is selected, the user can choose to (1) view the documentation, (2) choose an input file different than the default file specified in the database, or (3) run the code (typically from its own menu-driven DOS batch file). All engineering groups should have a utility like this. The home page and contents are shown below, along with the example of choosing the nozzle flow code TDKMS:

Select Category:

Select Code:

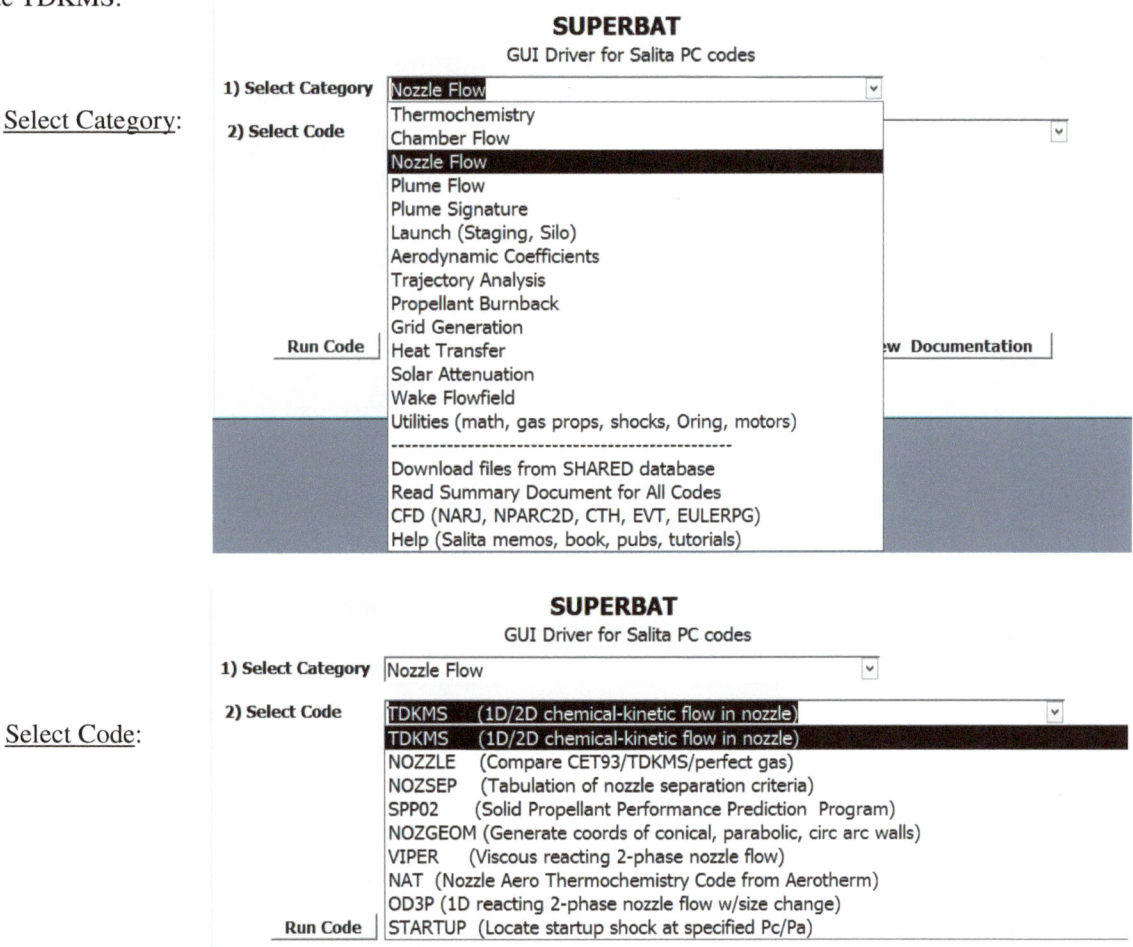

Data Base for Nozzle Flow (sometimes I have to look here to remind myself of the path to the selected code):

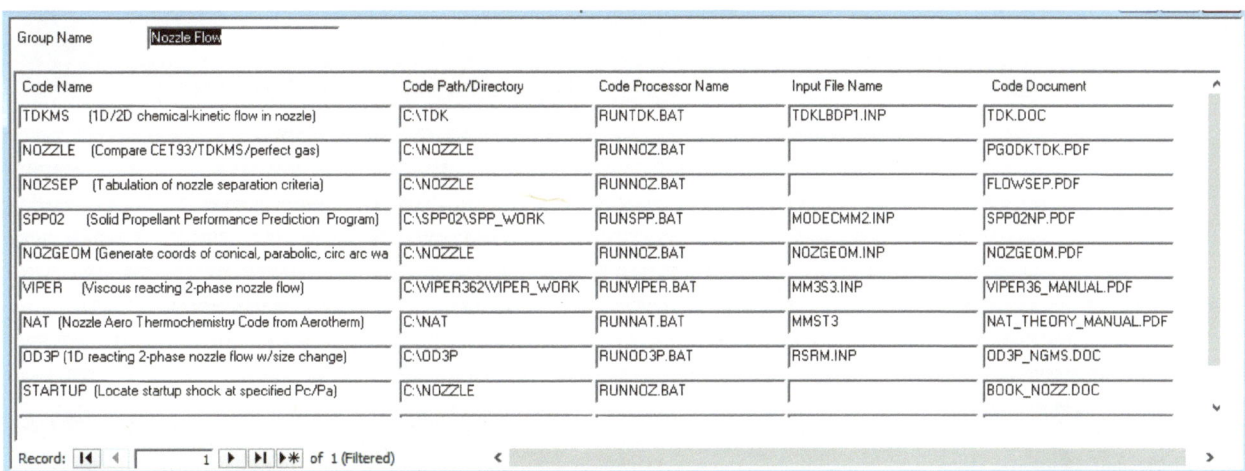

Utility Codes

During my nearly 40 years in the rocket business, I often needed to evaluate gas and thermochemical properties, mathematical functions, and parameters tabulated in book appendices or specialized reports like NACA 1135. Rather than having to search for them, and then to interpolate manually in tables for precise values, I decided to automate as many of these properties as possible. The result is my package of Utility Codes:

```
    ---------------------------------------
         BATCH FILE FOR SALITA'S UTILITY CODES
    ---------------------------------------

  CHOOSE UTILITY OPTION:
     0) QUIT
     1) Run GASP    to calculate gas properties (isentropic, emissivities, atmosphere, free-molecular)
     2) Run MATH    to calculate mathematical functions (Bessel, Log-Normal, Error Fcn, etc)
     3) Run THERMO  to calculate thermodynamic properties vs temperature and reaction
     4) Run SHOCKS  to calculate shock jump relations over wedge, cone, sphere, during nozzle startup
     5) Run PLUMES  to calculate near-vac plumes, viscous inert plume mixing, particle trajectories
```

The menu options for all five utilities are listed below:

GASP:
```
        CHOOSE A GAS-PROPERTY FUNCTION:
           0) Quit

           1) Run ISENTROP to evaluate isentropic props given Mach#, A/A*, or P/Pc

           2) Run PRANDMEY to calculate Mach number given Prandtl-Meyer angle or vice versa

           3) Run BLACK    to evaluate/plot black body radiancy given T(K), emissivity

           4) Run BANDMOD  to tabulate/plot bandmodels for gaseous species and soot
                               given BMPY.DAT, BANDMOD.SPU, OPT192.DAT, BANDMOD.O3
           5) Run KABS     to evaluate emissivity of mixture of gases (a la PLUMESIG)
                               given p, T, L, mole fractions
           6) Run HEMIS    to evaluate hemispherical emissivities of cylindrical gas/particle cloud
                               using closed-form solution (Shadlesky/Salita)
           7) Run ATMOS    to calculate p,T,density at specified altitude H
                               for 1976, SPF, or MCAT atmosphere
           8) Run FREEMOLE to calculate aero coefficients in free molecular flow
                               assuming specular isoenergetic reflection
           9) Run TRANPROP to calculate transport properties for gas mixture

           L) Run LENKEL   to calculate flow through close-clearance orifices
```

MATH:
```
        CHOOSE A MATH FUNCTION:
           0) QUIT
           1) Evaluate  BESSEL FCN:   calculate I0,I1,J0,J1,K0,K1(X) given X
           2) Evaluate  BESSEL INTEGRAL: exp(eta/cost) Io(eta) deta
           3) Evaluate  ELLIPTIC INTEGRALS of 1st and 2nd Kind
           4) Evaluate  ERROR FCN:   calculate ERF(X) given X or X given ERF
           5) Calculate a LOG-NORMAL size distribution given Dm,sig
                   or discrete diameters to simulate log-normal distribution
           6) Linearly INTERPOLATE is specified table
           7) Determine polynomial LEAST-SQUARE curve fit of data array POINTS.DAT
           8) COMPARE two files (or a table of file pairs) character-by-character
           9) Draw vehicle trajectory on MAP of world (from Suarez/Teuscher)
```

SHOCKS:
```
        CHOOSE SHOCK OPTION TO RUN:
           0) QUIT
           1) Run WEDGE     to calculate properties behind WEDGE shock
                               (specify gamma, Minf, half-angle)
           2) Run SCONE     to calculate shock layer on supersonic CONE
                               (specify gamma, Minf, half-angle)
           3) Run MN2IT     to calculate shock layer on SPHERE
                               (specify gamma, Minf)
           4) Run SHOKTUBE  to calculate unsteady properties in SHOCK TUBE
                               (need SHOKTUBE.INP)
           5) Run STANDS    to calculate location of standing shock in nozzle
                               (specify gamma, Pc/Patm, Ae/A*)
           6) Run STARTUP   to calculate nozzle startup shock given ignition transient
                               (need STARTUP.INP)
           7) Run ISENTROP  to calculate shock jump properties in perfect  flow
                               (better to run from GASP)
           8) Run JUMP      to calculate shock jump properties in reacting flow (need JUMP.INP)
                               or Chapman-Jouget conditions      (need CHAPMANJ.INP)
           9) Run BLAST     to calculate blast wave from TNT (White Sands, Brode, Kinney,
                               or Kingery/Bulmash (specify 2 of R,W,I,dp))
```

__THERMO__

```
                MAKE CHOICE:
                  0) QUIT

                  1) Run JANDOW   to process  thermochemical product species data bases
                                                 (JANDOW.DAT, NIST,DAT, \CFDRC\SPECIES.DAT)
                  2) Run CHON     to estimate product composition for CHON propellant
                                                 using closed-form solution
                  2) Run EQCON    to evaluate  equilibrium constant for arbitrary-format react
                                                 (specify T(K) and reaction in EQCON.INP)
                  3) Run CHEMCF   to estimate product composition for CHONClAl propellant
                                                 using closed-form soln (input=CHEMCF.INP)
                  4) Run PLUMIX   to predict   inert mixing of rocket exhaust with atmosphere
                                                 (need PLUMIX.INP)
                  5) Run AFTERB   to predict   plume afterburning using closed-form soln
                                                 (need AFTERB.INP)
                  6) Run MASSMOLE to convert   mole fractions to mass fractions or vice versa
                                                 (specify input fractions in MASSMOLE.INP)
                  7) Run MASSFRAC to convert   SPFIII mole fraction file \SPFIII\SPF3SAX.PLT
                                                 to mass fractions
                  8) Run QENBOM   to process quench-bomb residue data (need data file)

                  9) Run finite-rate chemistry solutions

                  T) Run TRANSP   to calculate transport properties
                                                 (need TRANS.INP)
```

Example of next level for JANDOW :

```
                CHOOSE YOUR JANDOW OPTION ...
                    1) GENERATE LIST OF ALL SPECIES
                    2) GENERATE LIST OF SPECIES CONTAINING SPECIFIED ELEMENTS
                    3) GENERATE TABLES OF Cp(T), S(T), H(T)-Hf
                         FOR SINGLE SPECIES TO BE LISTED BELOW
                    4) COMPUTE Cp, S, H-Hf AT SINGLE TEMPERATURE
```

__PLUMES__

```
              MAKE CHOICE FOR "PLUMES"
                  0) Quit
                  1) Run MOC       to predict near-vacuum plume using Method of Characteristics
                                              and particle trajectories using PLUMEPAR
                  2) Run ELLIP     to approximate MOC solution with elliptic contours
                                              and compare to DRAPHILL contours
                  3) Run DRAPHILL  to model axisymmetric far-field near-vacuum gas plume
                                              using Draper/Hill point-source model
                  4) Run PEARCE    to model particle trajectories thru near-vacuum plume
                                              using Pearce's point source model
                  5) Run SCARFP    to model near-vacuum plume from scarfed nozzle
                                              using Romine/Noble method
                  6) Run DONGRAY   to model mixing gaseous plume
                                              using integral method of Donaldson/Gray
                  7) Run PLUMIX    to model centerline decay of inert gaseous mixing plume
                                              as a function of fuel fraction FF
                  8) Run IMPINGE   to model impingement of elliptical plume on bodies
```

Example of next level for MOC:

```
                MAKE CHOICE FOR "MOC"
                   0) Return to main menu
                   1) Edit input  file "MOC.INP"
                   2) Run  MOC
                   3) Edit output file "MOC.OUT"
                   4) Plot flowfield properties   from "MOC.PLT"
                   5) Plot centerline Mach growth  from "MOC.AXI"
                   6) Plot radial Mach distribution from "MOCRAD.PLT"
                   6) Plot radial Mach distribution from "RADIAL.EXE"
                   7) Plot variables along boundary from "MOC.BND"
                   8) Plot properties along surface from "MOC.IMP"
                   9) Plot Mach dist on spheric cap from "MOC.PLT"
                   R) Edit boundary reflection file    "MOC.REF"
                   G) Browse User's Guide              "MOC.GID"
```

Examples of Derivation of Closed-Form Solutions   *(Warning: Math Ahead!)*

For the math students among you, I'm providing some detailed examples of how I derived some closed-form solutions.

Chamber Blowdown

Suppose we have a constant volume V containing ideal gas at high pressure p_o and temperature T_o at a time t_o when either an orifice suddenly opens or propellant in a rocket chamber burns out. The volume will begin to exhaust or "blow down" to atmospheric pressure p_{atm}. We wish to determine the resulting history of pressure $p(t)$ and temperature $T(t)$ in volume V.

The orifice (or nozzle throat) of area A* will remain choked for all but the final time increment of the blow-down process, i.e. until pressure ratio p/p_{atm} drops below $[(\gamma+1)/2]^{\gamma/(\gamma-1)}$, where γ is the ratio of specific heats (≈ 1.2). A closed-form solution describing this process can be obtained from the unsteady mass and energy conservation equations (5,6) on page 21.

For a blow-down process ($\dot{m}_{in}=0$) that is adiabatic ($\dot{Q}_{out}=0$) and choked ($\psi^*=1$), mass and energy conservation equations (5,6) reduce with specific gas constant R to

$$\dot{m}_{out} = \dot{m}_{nozz} = \frac{pA^*}{C^*(T)} \qquad \text{where} \quad C^*(T) = \left[\frac{\beta^{\beta/\alpha}}{\gamma}RT\right]^{1/2} = C^*(T_o)\, \bar{T}^{1/2}$$

$$\frac{dp}{dt} = \frac{RT}{V}[-\dot{m}_{out}] + \frac{p}{T}\frac{dT}{dt} \quad = \left[-B\bar{T}^{-1/2} + \frac{1}{T}\frac{dT}{dt}\right]p \qquad \text{where} \quad B = \frac{RT_oA^*}{VC^*(T_o)} \qquad \bar{T} = \frac{T}{T_o}$$

$$\frac{dT}{dt} = \frac{RT}{pV}[\gamma(-\dot{m}_{out}T) + \dot{m}_{out}T] = -(\gamma-1)B\bar{T}^{3/2}T_o \qquad \alpha = \frac{\gamma-1}{2} \qquad \beta = \frac{\gamma+1}{2}$$

Integrating the temperature equation from time t=0 to time t yields

$$\int \bar{T}^{-3/2}d\bar{T} = -2\left(\bar{T}^{-1/2}-1\right) = -(\gamma-1)\,B\,t$$

or

$$\boxed{T(t) = \frac{T_o}{(1+\omega t)^2}} \qquad \text{where} \quad \omega = \alpha B$$

Plugging dT/dt into the pressure equation yields

$$\frac{1}{p}\frac{dp}{dt} = -\gamma B\bar{T}^{1/2} = -\gamma B\frac{1}{1+\omega t}$$

or

$$\ln\frac{p}{p_o} = -\gamma B\frac{\ln(1+\omega t)}{\omega} = \ln[(1+\omega t)^{-\gamma B/\omega}]$$

or

$$\boxed{p(t) = \frac{p_o}{(1+\omega t)^{2\gamma/(\gamma-1)}}}$$

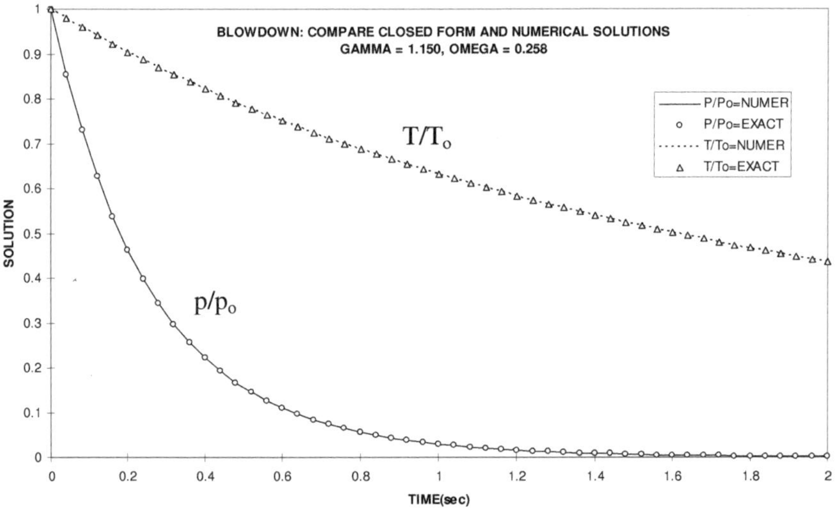

BLOWDOWN: COMPARE CLOSED FORM AND NUMERICAL SOLUTIONS
GAMMA = 1.150, OMEGA = 0.258

Agreement of the closed-form solution with a numerical integration of the differential equations is shown here to be excellent. Note that chamber pressure drops much faster than chamber temperature.

Solution of 1D Heat Conduction Equation

It is not uncommon for (a) an analyst to model a heat conduction problem numerically when an exact analytical solution to it or a closely related problem exists, and (b) the resulting numerical solution to be inaccurate due to a poor choice of nodal (grid) spacing. Consequently, it is useful to have a wide range of exact closed-form solutions available to provide sanity checks and validation methods for the numerical solutions to related problems. To do so, it is necessary to derive closed-form solutions to the one-dimensional heat conduction equation

$$\boxed{\frac{\partial T}{\partial t} = \alpha\frac{\partial^2 T}{\partial y^2} + \left[\alpha\left(K \pm \frac{\varepsilon}{r}\right)\right]\frac{\partial T}{\partial y}} \qquad\qquad\qquad (\text{HC-1})$$

where $\alpha = k/\rho C_p$ is the constant thermal diffusivity, $K = \partial k/\partial y = (\partial k/\partial T)(\partial T/\partial y)$ accounts for the variability of thermal conductivity k with temperature T, ρ is the mass density and $C_p(T)$ the specific heat capacity of the material; t is time and y is depth from the material surface; $\varepsilon=0$ for planar surfaces, 1 for cylindrical surfaces, and 2 for spherical surfaces (the sign on ε is positive when heating the interior of a hollow body, but negative when heating from the outside because r and y would be in opposite directions).

A closed-form solution to eq(HC-1) is possible for constant material properties and heat flux parameters (next page).

Solutions to eq(HC-1) for <u>constant α and k</u> (K=0) have been obtained for heating of planar surfaces (ε=0) using Laplace transformation and convolution (see Ref 78 or most applied-math text books) subject to two types of boundary condition on heat flux $\dot{q}(t)=-k(\partial T/\partial y)_w$:

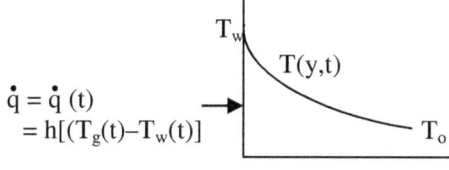

(a) $\dot{q}(t)$ = specified constant

(b) $\dot{q}(t) = h[(T_g(t)-T_w(t)]$

$\dot{q} = \dot{q}(t)$
$= h[(T_g(t)-T_w(t)]$

where h is the constant heat transfer coefficient and T_g is the constant temperature of the gas adjacent to the surface whose temperature is $T_w(t)$. For both boundary conditions, the initial uniform temperature of the material is T_o.

Heating of a Solid Material

Eq(HC-1) is often solved by applying the Laplace Transformation to convert it into an ordinary differential equation for the transformed temperature $\bar{T}(y,s)$ dependent on y alone. Applying the Laplace transformation $\boldsymbol{L}$ to eq(HC-1) and its boundary conditions yields

$$\boldsymbol{L}\{\frac{\partial T}{\partial t} = \alpha \frac{\partial^2 T}{\partial y^2}\} \quad \rightarrow \quad s\bar{T} - T_o = \alpha \frac{d^2\bar{T}}{dy^2} \tag{HC-2}$$

whose solution is

$$\bar{T}(y,s) = \frac{T_o}{s} + Ae^{my} + Be^{-my} \quad \text{where} \quad m = (\frac{s}{\alpha})^{1/2} \tag{HC-3}$$

subject to

$$\boldsymbol{L}\{T(y=\infty,t) = T_o\} \quad \rightarrow \quad \bar{T}(\infty,s) = \frac{T_o}{s} \quad \rightarrow \quad A = 0$$

$$\boldsymbol{L}\{-k(\frac{\partial T}{\partial y})_{y=0} = \dot{q}(t)\} \quad \rightarrow \quad -k(\frac{d\bar{T}}{dy})_{y=0} = \dot{q}(s) \quad \rightarrow \quad B = \frac{\dot{q}(s)}{mk}$$

The solution in the physical plane is now determined by first applying the inverse Laplace transform $\boldsymbol{L}^{-1}$ to eq(HC-3):

$$T(y,t) = \boldsymbol{L}^{-1}\{\frac{T_o}{s}\} + \boldsymbol{L}^{-1}\{\dot{q}_o(s)\frac{e^{-my}}{mk}\} = T_o + \frac{\alpha^{1/2}}{k}\boldsymbol{L}^{-1}\{f_1(s)\ f_2(s)\} \tag{HC-4}$$

where

$$f_1(s) = \dot{q}(s) \qquad \boldsymbol{L}^{-1}[f_1(s)] = \dot{q}(t)$$

$$f_2(s) = \frac{e^{-(s/\alpha)^{1/2}y}}{s^{1/2}} \qquad \boldsymbol{L}^{-1}[f_2(s)] = F_2(t) = \frac{1}{(\pi t)^{1/2}}\exp[-\frac{y^2}{4\alpha t}] \quad \text{(Ref 78)}$$

Application of the Convolution Theorem yields

$$T(y,t) = T_o + \frac{\alpha^{1/2}}{k}\int_0^t \dot{q}(t-\tau)F_2(\tau)d\tau = T_o + \frac{\alpha^{1/2}}{k}\int_0^t \dot{q}(t-\tau)\frac{\exp(-y^2/4\alpha t)}{(\pi t)^{1/2}}d\tau \tag{HC-5}$$

The integral in eq(HC-5) is easily evaluated for <u>constant heat flux</u> $\dot{q}=\dot{q}_o$ into a thermally-thick planar solid (i.e. where the thermal wave penetrates to a depth y<L where L is the physical depth of the solid):

$$T(y,t) = T_o + \frac{\dot{q}_o}{k}F(y,\alpha t) \qquad \text{where} \quad F = (\frac{4\alpha t}{\pi})^{1/2}\exp(-Z^2) - y\ \text{erfc}(Z) \tag{HC-6a}$$

$$T_w(t) = T_o + \frac{\dot{q}_o}{k}(\frac{4\alpha t}{\pi})^{1/2} \qquad Z = \frac{y}{(4\alpha t)^{1/2}} \tag{HC-6b}$$

T_w is the wall (surface, y=0) temperature, and erfc = 1–erf is the complementary error function (Ref 78).

Alternately, for <u>constant heat transfer coefficient h and gas temperature T_g</u> in $\dot{q}=h(T_g-T_w)$:

$$T(y,t) = T_o + (T_g-T_o)\Phi(Y,Z) \tag{HC-7a}$$
$$T_w(t) = T_o + (T_g-T_o)[1-\exp(Y^2)\text{erfc}(Y)] \tag{HC-7b}$$

where $\Phi = \text{erfc}(Z) - e^{(Y+2Z)Y}\text{erfc}(Y+Z)$ $Y = H(\alpha t)^{1/2}$ $H = \frac{h}{k}$

Eq(HC-6b), in conjunction with laser ignition data, has been used to determine the ignition temperature of propellant (for example, see page 77).

Eqs (HC-6,7) have been used to validate numerical heat conduction codes.

Anatomy of a Problem and Its Solution: Modeling O-Ring Deformation/Activation

This discussion is meant to provide a detailed description of the steps I had to take to solve one problem. The final solution for the Shuttle booster is documented in Ref 27, but the process required to obtain that solution is described below.

As discussed on page 29, I returned to Thiokol from the Challenger accident investigation with the knowledge that the failed joint and O-rings had been at a temperature far lower than previously known. We needed to determine quickly how that low temperature affected the ability of the O-rings to seal upon motor ignition. The world was watching, and we needed to redesign the booster joints. Unfortunately, no analytical model existed at that time, not even from the largest O-ring manufacturer Parker Seals, whose Handbook only contained empirical (measured) data.

Since I had previously done the modeling of the joint pressurization and O-ring erosion, I suggested to my management that I try to develop an O-ring deformation/activation model. They agreed. The problem was that the model required structural analysis, whereas I was a fluid mechanics specialist. My last involvement with structural analysis was as an undergraduate senior at RPI more than 20 years earlier. However, if I could identify an appropriate set of simplified equations, maybe I could generate a reasonable approximate solution, what is termed in industry as "an engineering solution" that gives sufficiently-accurate answers, even though the model is not formally and precisely correct.

So I went to our in-house specialist in structural mechanics, Dr. Ron Webster. I asked if he could define for me a simple model of material deformation that I could rapidly couple to my model of joint pressurization. He said that a linear "small-deformation" model was the only "simple" option, but that as soon as the deformation becomes obvious, the solution would become formally invalid. Even though an O-ring deforms a lot, I had no choice. We needed a viable model within weeks.

I started by approximating the undeformed circular O-ring cross-section by a set of N=36 triangular "finite elements" (like pie slices), all with common node at the center. A diagram of the cross-section of the booster O-ring of undeformed diameter D_o=280 mils in the joint groove is shown here, and defines the initial coordinates of the 37 nodes at the element corners.

> Lesson #1: This figure demonstrates the value of storing and saving input files for future use. I regenerated this clean figure today by running ORINGDEF using the input listed below, 28 years after I first generated it for Ref 27. Otherwise, I would have had to scan the Journal article to generate at best a hazy version.

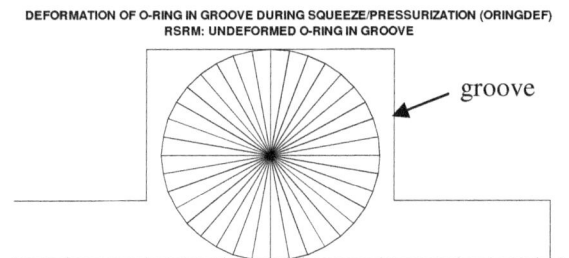

DEFORMATION OF O-RING IN GROOVE DURING SQUEEZE/PRESSURIZATION (ORINGDEF)
RSRM: UNDEFORMED O-RING IN GROOVE

groove

```
RSRM: UNDEFORMED O-RING IN GROOVE
&FLAG  IPROB=1, NELEM=36, MULT=0, IDEBUG=0, IPLOT=1,            &END
&GEOM  DRING=0.280, GAP=0.080, HEIGHT=0.200, WIDTH=0.317,
       RJOINT=1000.,                                            &END
&OPTS  TEMPF=-1780., POIS=0.495, ADT=0.0, PEXT=0., CFRIC=1.0,   &END
&INCR                                                           &END
```

Next we defined the components of strain for each node (strain is a measure of the spatial rate of nodal displacement), and the related components of stress from Hooke's Law for a uniform (homogeneous and isotropic) and incompressible material. The only cause of the stress was assumed due to gas pressure acting on the surface of each element. The effects of material stretch and thermal contraction $\alpha\Delta T$ were added. The resulting set of equations defining static equilibrium of the nodes of this single element was then written as

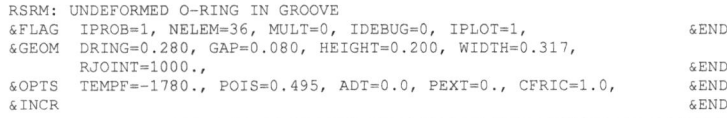

$$[K]\{q\} = \{f\}$$

where [K] is the elemental stiffness matrix, {q} is the matrix of nodal displacements, and {f} is the matrix of edge forces. This matrix equation is the same as that for a "**spring**". The equations for all N elements were combined to generate a set of "global static equilibrium equations" of the same form, where [K] would be the global stiffness matrix.

The boundary condition for a static O-ring is that when pressurized, no nodes can pass through the walls of the groove. This condition is achieved iteratively:
1) Fix one node of the O-ring at one surface of the groove, apply pressure.
2) Force the node that has displaced furthest past the wall, back to the wall.
3) Iterate until no nodes pass through the wall.

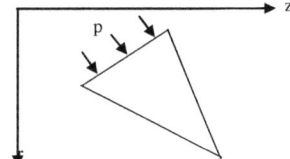

Fixed Point

Must push back to wall

An example of the first step is shown here. A pressure of 1000 psi was applied uniformly on the lower surface of the O-ring. However, on my first attempt during the development of the computer code, the O-ring deformed asymmetrically to one side. That shouldn't have happened.

> Lesson #2: In order to find my coding bug, I needed to do a hand calculation. That would take too long for 36 elements, so I tried it for only four elements (see page 27). There didn't seem to be an obvious coding error, but I eventually realized that the code needed to be "double-precisioned": the truncation error when using only single-precision was enough for a pressure force at 1000 psi to totally skew the displacements asymmetrically. This fixed the problem.

The resulting code **ORINGDEF** was then ready to be applied to a wide variety of validations and applications!

Static Validation Studies

1) The static O-ring was <u>squeezed</u> different amounts from the bottom (sealing) surface, and the predicted footprints b/D_o were shown to agree well with values measured by Molari:

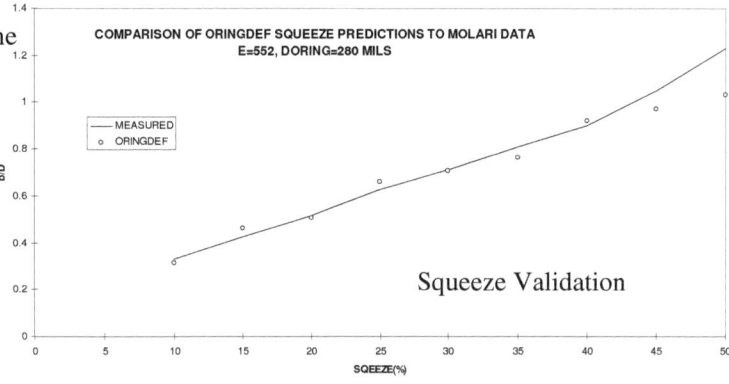

ORINGDEF: 25% SQUEEZE

Footprint b

Squeeze Validation

2) Predicted depths of O-ring <u>extrusion</u> into three different joint gaps at both 25°F and 75°F were shown to agree well with measurements made at 1000 psi at Thiokol.

3) The predicted reduction in O-ring cross-section due to <u>stretching</u> its hoop diameter (which occurs during installation in the booster joint groove) was shown to agree with measured data from the Parker Handbook.

4) The predicted effect of <u>thermal contraction</u> on O-ring shrinkage was shown to agree with predictions from the structural analysis code ABAQUS and Parker Handbook.

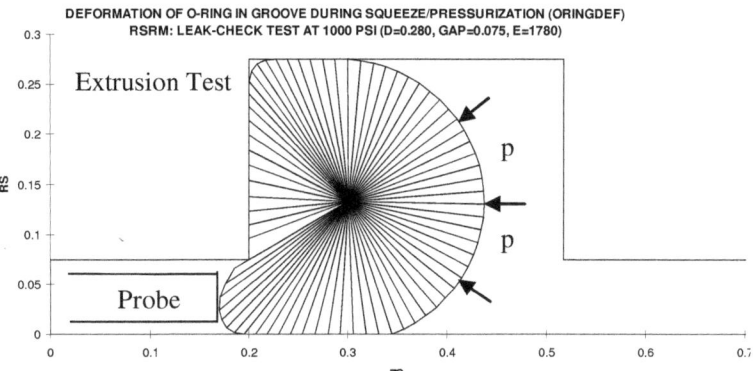

Extrusion Test

Probe

Dynamic Validation Studies

All Shuttle booster joints were "leak-checked" before launch by pressurizing the volume between the primary and secondary O-rings. This would push the secondary O-ring aftward into its sealing position, but would push the primary O-ring forward (like the extrusion test shown above). When the motor ignites, the gas would pressurize from the forward side and the squeezed O-ring would have to first overcome <u>static friction</u> before sliding into its sealing position at the aft side.

In order to determine the coefficient of static friction μ of the greased O-ring, I formulated and ran a series of "table-top" tests. A circular groove was machined in the bottom of a steel plate, and a greased O-ring was inserted in the groove. A known weight of additional steel was placed on the top of the plate, and the plate was tilted until the block began to slide at an angle θ_{cr}. The component of gravitational force tangent to the sliding surface then resulted in $\mu = \tan \theta_{cr}$.

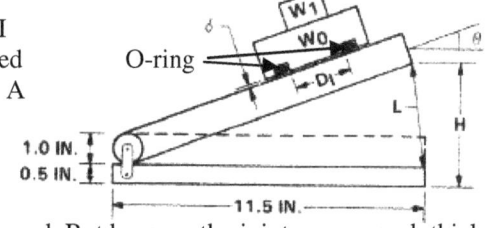

During motor ignition, the pressurization of the chamber caused the case to expand. But because the joints were much thicker than the case wall, they acted like a belt that limited the expansion there. This caused the <u>joint gap to widen rapidly</u> ("rotate") during the 600 ms pressurization process. The O-ring would thus have to expand radially as fast as the gap opened, or the joint would leak. At colder temperatures, the reduced "resiliency" of the O-ring would slow its ability to expand radially; this was modeled here by introducing a rate term (linear "**dashpot**") whose force is assumed proportional to stiffness:

$$[K]\{q\} + C[K]\{\dot{q}\} = F$$

Spring   Dashpot

Joint Rotation and Gap Growth
(Greatly Exaggerated)

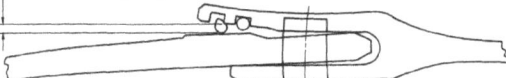

A curve-fit for the "resiliency coefficient" C was determined by matching a large quantity of experimental data obtained during the accident investigation. The resulting time-dependent spring-dashpot equations were marched in ORINGDEF to determine the history of O-ring deformation and activation at both 25°F and 75°F, and is shown on page 30.

<u>Lesson #3</u>: This linear small-deformation model worked well due to the dominance of the groove walls over the material properties in constraining the deformation of the O-ring. This engineering success totally astounded the structural analysts.

<u>Lesson #4</u>: Sometimes politics gets in the way. I believe that my conference paper documenting these studies (Ref 27) should have won Best Paper of the Year; it had a clean analytical solution that included a multitude of real factors (squeeze, compression-set, stretch, thermal contraction, friction, etc), validation to tons of experimental data, and world-wide interest. The conference paper after mine had been canceled, and the huge audience insisted that I use that available time to continue my presentation. However, the chairman of my conference session was from NASA, and he didn't even recommend it for an award, I assume because of NASA's sensitivity to the O-ring issue. You read the article and decide.

My Book

I decided to upgrade my Short Course into an electronic book that would be more complete, have better graphics, and be easily supplied to my Northrop colleagues, as well as to the U.S. rocket industry. The book is provided in three formats: text (550 pages), slides (900), and Highlights (101 pages). I submitted a version to the Pentagon in 2008 for approval for distribution. They sent it out to several organizations for review. The Air Force Rocket Lab (AFRL) at first felt it should be ITAR restricted (limited to the U.S. Government and its contractors) since they believed "anyone could design a rocket using this book" (that is silly, since there is no material in the book about many rocket subjects like structural dynamics or guidance and control). The Pentagon officer overseeing the review process agreed with me, and sent the book to NASA Headquarters and the DOD. After I made some required corrections, it received approval for public release (DOD Ref 10-S-0085).

The book is titled <u>Basic Analytical and Numerical Methods for Propulsion and Aerodynamic Analysis of Solid Propellant Rockets</u>. I believe it to be the most complete book available on these subjects. However, since I don't want the book to go to Iran, North Korea, or even China, I have been distributing it as if it is classified as ITAR Class 3: available only to the U.S. Government and its Contractors, and foreign organizations friendly to the U.S. (currently only ESA and Israel). It has been provided to about 20 U.S. organizations (Northrop Grumman, Lockheed-Martin Denver, Orbital, ATK, Raytheon, Aerojet, Aerospace Corporation, NASA (JSC, MSFC, Headquarters), AFRL, Redstone Arsenal, China Lake, SSI, SRA, CRAFT, …).

Basic Analytical and Numerical Methods for Propulsion and Aerodynamic Analysis of Solid-Propellant Rockets

By Dr. Mark Salita

Equations, Solution Methods, and Applications for:

Characteristics of Solid Propellants
Ignition of Solid Propellants
Combustion of Solid or Liquid Propellants
Chamber Flow in Solid Rockets
Nozzle Flow in Solid or Liquid Rockets
Heat Conduction and Material Ablation in Rockets
Launch Issues for Solid or Liquid Rockets
Exhaust Plumes from Solid or Liquid Rockets
Vehicle Aerodynamics for Solid or Liquid Rockets
Vehicle Trajectory Analysis and Optimization
Appendices
Index
Homework Problems

Space Shuttle

Constellation Concept

References

1) Salita, M., <u>Basic Analytical and Numerical Methods for Propulsion and Aerodynamic Analysis of Solid-Propellant Rockets</u>, self-published electronic book, revised August 2011.

2) Salita, M., "Unanticipated Problems and Misunderstood Phenomena in and Around Solid Rockets", AIAA-2011-5956, 30th Joint Propulsion Conference (San Diego), August 2011.

3) Salita, M., "Shuttle Disasters: A Common Cause?", Aerospace America, March 2004, pp41-43.

4) Salita, M., "Deficiencies and Requirements in Modeling of Slag Generation in Solid Rocket Motors", Journal of Propulsion and Power, Vol.11 No.1, January-February 1995.

5) Salita, M., "Exact Bookkeeping of I_{sp} Penalties", JANNAF PSS 11th Meeting Minutes, February 1978, pp 239-254.

6) Salita, M., "New Approaches for Calculating 1D and 2D Two-Phase Flow", JANNAF PSS 11th Meeting Minutes, February 7, 1978, pp 215-238.

7) Salita, M., "Exact Bookkeeping Study of I_{sp} Penalties: Comparison of AFRPL (SPP) and Thiokol (TPP) Systems", JANNAF PSS 12th Meeting Minutes, January 1979, pp 187-192.

8) Salita, M., "A Proposed Reaction Set for Kinetic Calculations in Solid Propellant Rocket Nozzles", CPIA 321, JANNAF PSS 13th Meeting Minutes, July 1980, pp 49-52.

9) Salita, M., "Prediction of Nozzle Boundary Layers Using Aerotherm's MEIT Code", CPIA 321, JANNAF PSS 13th Meeting Minutes, July 1980, pp 293-312.

10) Salita, M., "Examination of Recent Models for Particle Drag in Nozzle Flows", CPIA 321, JANNAF PSS 13th Meeting Minutes, July 1980, pp 347-350.

11) Salita, M., "Analysis of Ignition Transients in Solid Rocket Motors", CPIA 344, JANNAF PSS 14th Meeting Minutes, February 1981, pp 205-222.

12) Salita, M., "Use of SOLA to Predict Unsteady Incompressible Viscous Flow in Burning Grain Slots", CPIA 355, JANNAF PSS 15th Meeting Minutes, March 1982, pp 205-230.

13) Salita, M., "A Simple Approximate Method for Predicting Equilibrium Chemical Composition in Rocket Motor Flows", CPIA 382, JANNAF PSS 16th Meeting Minutes, March 1983, pp 181-206.

14) Salita, M., "Some Numerical and Physical Experiments to Simulate Propellant Casting", CPIA 382, JANNAF PSS 16th Meeting Minutes, March 1983, pp 271-302.

15) Salita, M., "Some Effects of Particle Size Averaging on Rocket Motor Performance Prediction", CPIA 417, JANNAF PSS 17th Meeting Minutes, March 1984, pp 95-114.

16) Salita, M., "Comparison of Several Finite Difference Schemes for Solving Transient Inviscid Flows with Shock Waves", AIAA-85-1126, 21st Joint Propulsion Conference (Monterey), July 1985.

17) Salita, M., "Observations of Several Resin Creep/Casting Experiments", Thiokol Report 2814-81-M156, Nov 6, 1981.

18) Salita, M., "Analysis of the Flowfield During Casting of Uncured Solid Propellants", Thiokol Report 2814-82-M047, February 22, 1982.

19) Salita, M., "Geometric Equations for Planar Burnback of a Star-Shaped Grain", Thiokol memo 2814-80-M150, December 10, 1980. Also "Planar Burn Back of a Planar Star", TRW memo F321.MS.97-005, April 17, 1997.

20) Salita, M., "Rocket Propellant Ignition: A Workshop Report", CPIA 432, 22nd JANNAF Combustion Meeting (JPL), October 1985, pp 65-86.

21) Salita, M., "Modification of VOLFIL to Model the Ignition of Multiple Wet or Dry Grains", Thiokol Report 2814-83-M100, August 15, 1983.

22) Salita, M., "A Simple Physical and Mathematical Model of the Operation of Gas-Bag Inflator", Thiokol Report, 2814-80-M043, April 10, 1980.

23) Salita, M., "Predicted Ingestion of Humidity Into the Gas-Bag Inflator", Morton-Thiokol Report 2814-FY1985-M186, May 7, 1985.

24) Salita, M., "Volume-Filling Simulation of Interstage Cavity Pressurization", Thiokol Report 2850-FY91-M236, May 21, 1991.

25) Salita, M., "Verification of Spatial and Temporal Pressure Distributions in Segmented Solid Rocket Motors", AIAA 89-0298, 27th Aerospace Sciences Meeting (Reno), January 10, 1989.

26) Salita, M., "Description of the Shock Waves Generated During the Initial Ignition Transient of the Shuttle SRBs", Thiokol memo 2814-81-M095, July 28, 1981.

27) Salita, M., "A Simple Model of O-Ring Deformation and Activation During Squeeze and Pressurization", Journal of Propulsion and Power, Vol.6 No.6, November 1988, pp497-511.

28) Salita, M., "Closed-Form Analytical Solutions for Fluid-Mechanical, Thermochemical, and Thermal Processes in Solid Rocket Motors", AIAA 98-3965, 34th Joint Propulsion Conference (Cleveland), July 14, 1998.

29) Salita, M., "Understanding ACE Charring Thermochemistry Using the NASA-Lewis Thermochemistry Code or the 14-Species Simplified Model SIMPACE", Thiokol TWR-40314, November 14, 1991.

30) Salita, M., "Quench Bomb Investigation of Al_2O_3 Formation from Solid Rocket Propellants (Part II): Analysis of Data", CPIA 498 (Vol.I), 25th JANNAF Combustion Meeting (MSFC), October 28, 88.

31) Salita, M., "Use of Water and Mercury Droplets to Simulate Al_2O_3 Collision/Coalescence in Solid Rocket Motors", Journal of Propulsion and Power, Vol.7 No.4, July 1991, pp505-512.

32) Salita, M., "Implementation and Validation of the One-Dimension Gas /Particle Code OD3P", CPIA 529 (Vol.II), 26th JANNAF Combustion Meeting (JPL), October 1989, pp69-82.

33) Salita, M., "Survey of Recent Al_2O_3 Droplet Size Data in Solid Rocket Chambers, Nozzles, and Plumes", CPIA 620 Vol.I, 31st JANNAF Combustion Meeting (Sunnyvale), October 1994.

34) Sambamurthi, J., "Al_2O_3 Collection and Sizing from Solid Rocket Motor Plumes", J. Propulsion and Power, Vol.12 No.3, June 1996, pp598-604.

35) Salita, M., "Interaction of Liquid Metal Droplets with Surfaces (Part I): Equilibrium Droplet Shapes", Thiokol Report 2850-FY92-M117, January 8, 1992.

36) Salita, M, "Emissivity of Gas/Particle Clouds in Chambers of Solid Rocket Motors", CPIA 557 (Vol.I), 27th JANNAF Combustion Meeting (Cheyenne), November 1990.

37) Salita, M., "Characterization of the Al_2O_3 Particulate Formed During Combustion of SRB Propellant (TP-H1148)", Thiokol TWR-18456, September 13, 1988.

38) Golafshani, M., Loh, H.T., Salita, M., Pratt, D., Smith-Kent, R., and Abel, R., "Prediction of Slag Accumulation in SICBM Static and Flight Motors", Thiokol TWR-40259, September 12, 1990.

39) Salita, M., "Flow Modeling and Computational Methods for Classical Hybrid Propulsion", Lecture Series on Hybrid Propulsion, 29th Aerospace Sciences Meeting (Reno), January 1991.

40) Salita, M., "Comparison of Four Boundary Layer Solutions for Fuel Regression Rate in Classical Hybrid Rocket Motors", AIAA-91-2520, 27th Joint Propulsion Conference (Sacramento), June 25, 1991.

41) Salita, M., "Automated PC Plotting with MS/EXCEL", TRW Report F321.MS.98-002, February 16, 1998.

42) Salita, M., "EGG: A Simple Elliptic Grid Generator for Body-Fitting 2D Internal Flows", TRW Report F757.37.93-001, July 12, 1993.

43) Salita, M., "MAGG: Multiblock Algebraic Grid Generator", TRW Report F321.MS.95-004, December 20, 1995.

44) Salita, M., "VORSTREM: A Simple and Versatile 2D Viscous Solver for Chamber Flow in Solid Rocket Motors", AIAA-94-2779, 30th Joint Propulsion Conference (Indianapolis), June 27, 1994.

45) Salita, M., "Predicted Slag Deposition Histories in Eight Solid Rocket Motors Using the CFD Model 'EVT'", 31st Joint Propulsion Conference (San Diego), July 11, 1995.

46) Salita, M., "Recent Advances in Modeling Slag Generation in Solid Rocket Motors", AIAA Large Booster Conference (Monterey), September 1994.

47) Salita, M., "Modeling of Slag Generation in Solid Rocket Motors: Chairman's Workshop Report", CPIA 620, 31st JANNAF Combustion Meeting (Sunnyvale), October 1994.

48) Mulrooney, M., "An Assessment of the Role of SRMs in the Generation of Orbital Debris", NASA TP-2007-213738, February 2007.

49) Salita, M., "Implementation at TRW of the Thiokol SRM Ignition Code SHARPIT", TRW Report F321.MS.95-003, August 9, 1995.

50) Salita, M., "Rocket Propellant Ignition: A Workshop Report", CPIA 432 22nd JANNAF Combustion Meeting (JPL), October 1985, pp65-86.

51) Salita, M., "Modern SRM Ignition Transient Modeling (Part I): Introduction and Physical Models", AIAA 2001-3443, 37th Joint Propulsion Conference (Salt Lake City), July 9, 2001.

52) Salita, M., "Reaction Screening Study for Minuteman Silo IOP", TRW Report F321.MS.97-004, March 20, 1997.

53) Salita, M., and Glatt, L., "Modeling of Ignition Overpressure in Minuteman Silos", AIAA 97-2720, 33rd Joint Propulsion Conference (Seattle), July 8, 1997.

54) Salita, M., "Flowfield and Sideload Modeling of the Staging Event: A Workshop Report", CPIA/JANNAF, July 2002.

55) Salita, M., "Automated Procedure to Cycle 1D CMA92 to Simulate 2D Heat Conduction and Ablation", Northrop Grumman Report PCFT08.MS.07-003, June 18, 2007.

56) Salita, M., "A User-Friendly Version of the Heating and Ablation Code CMA92", Northrop Grumman Report PCFT08.MS.06-012, September 11, 2006.

57) Salita, M., "Implementation at TRW of the JANNAF Standardized Plume Flowfield Code SPF-III Version 4.0", TRW Report F321.MS.98-005, July 27, 1998.

58) Salita, M., "A User-Friendly Version of ZEUS for Supersonic Tactical Missiles with Fins", TRW Report F321.MS.97-011, December 4, 1997.

59) Salita, M., "Evaluation of CFD Code CFD-ACE for Rocket Applications", AIAA 2000-3186, 35th Joint Propulsion Conference (Huntsville), July 17, 2000.

60) Salita, M., "Supersonic Flow Past a Cone", TRW Report F321.MS.97-005, June 13, 1997.

61) Salita, M., "NEWTON: A Simple Code to Estimate Aerodynamic Coefficients for Supersonic Vehicles With and Without Fins", Northrop Grumman Report 2ETM.MS.03-013, October 24, 2003.

62) Salita, M., "Use of MN2IT as a Blunt-Nose Start-Plane Generator for ZEUS97", Northrop Grumman Report, 2ETM.MS.01-010, September 10, 2001.

63) Salita, M., "Installation of Aeroprediction Code DATCOM07 at NGMS", Northrop Grumman Report, PCFT08.MS.07-008, November 20, 2007.

64) Salita, M., "Predicted Vaporization of Accidentally-Cast Lead Pellets in Minuteman Motors", Northrop Grumman Report 2ETM.MS.03-005, March 17, 2003. Also Addendum memo 2ETM.MS.03-007, May 1, 2003.

65) Salita, M., and Bennett, D., "Model of Minuteman Launcher Closure Operation", Northrop Grumman Report 2ETM.MS.05-004, September 5, 2005.

66) Salita, M., "Enhanced Model of Minuteman Launcher Closure Operation", Northrop Grumman report PCFT08.MS.07-007, Sept 21, 2007.

67) Salita, M., "Simulation of Operation of TVC Hot Gas Relief Valve on Minuteman III Stage 2", Northop Grumman Report PCFT08.MS.07-002, April 3, 2007.

68) Salita, M., "Variation in the Performance of Mag/Teflon Igniters for Standard Missile", Thiokol Report 2814-82-M198, December 10, 1982.

69) Salita, M., "Performance Model of Black Brant V Ignition Using B/KNO3 Igniter", Northrop Grumman Report MEC-08-10165, September 2007.

70) Salita, M., "CTH Simulation of Detonation of LOX/LH2 Tank on CEV Stage 2", Northrop Grumman Mission Systems, Report PCFT08.MS.06-005, June 15, 2006.

71) Salita, M., "User-Friendly Version of the Russian Plume Codes NARJ/PRCJ", Report for AEDC, August 9, 2011.

72) Salita, M., "Validation of NARJ Nozzle and Plume Flowfield Simulator and PRCJ Plume Radiation Codes", Report for AEDC, April 13, 2012.

73) Salita, M., "Simple but Approximate Solutions for Radiant Intensity of Uniform Cylindrical Plumes", JANNAF 34th EPSS Conference, December 8, 2014.

74) Caveny, L., Tietz, D.E., and Salita, M., "Boost Phase Intercept of Iranian and North Korean Missiles Using Interceptors from UAVs", AIAA Defense 2015 Forum, JHU/APL, Laurel, MD, 10-12 March 2015.

75) Caveny, L.H., Tietz, D.E., and Salita, M., "UAV-Based Boost Phase Intercept (BPI) System to Defeat Rogue Nation ICBMs", Space and Missile Defense Symposium, Huntsville, AL, August 11, 2015.

76) Bate, R.R., Mueller, D.D., White, J.E., Fundamentals of Astrodynamics, Dover, 1971, pp 309ff.

77) Salita, M., "Comparison of Several Finite Difference Schemes for Solving Transient Inviscid Flows With Shock Waves", AIAA 85-1126, 21st Joint Propulsion Conference (Monterey), July 7, 1985.

78) Abramowitz, M., and Stegun, I, Handbook of Mathematical Functions, NBS Applied Mathematics Series #55, 1966.

\MARK\Rocket_Memoirs.doc

Pegasus XL Air Launch

United States Air Force

Certificate of Appreciation

to

MR. MARK SALITA

For your outstanding participation in the Propulsion Group of the Combined Contractor and Government Pegasus XL Failure Investigation Team. Your dedication, expertise and hard work during many hours in support of the investigation team were both instrumental and essential to a prompt, decisive and economical failure findings briefing that were well received at both Space and Missile Systems Division and the Air Staff. The investigation team is grateful for your efforts. Thank you, again, for a job supremely well done.

15 SEP 1994
Date

RICHARD C. POCH, Col, USAF
Director,
Operations and Support Directorate